WASTE AND WASTE MANAGEMENT

ENZYMES IN VALUE-ADDITION OF WASTES

Waste and Waste Management

Additional books in this series can be found on Nova's website under the Series tab.

Additional e-books in this series can be found on Nova's website under the e-book tab.

WASTE AND WASTE MANAGEMENT

ENZYMES IN VALUE-ADDITION OF WASTES

SATINDER KAUR BRAR
AND
MAUSAM VERMA
EDITORS

nova publishers
New York

For permission to use material from this book please contact us:
Telephone 631-231-7269; Fax 631-231-8175
Web Site: http://www.novapublishers.com

Additional color graphics may be available in the e-book version of this book.

Library of Congress Cataloging-in-Publication Data

ISBN: 978-1-62948-968-1

Published by Nova Science Publishers, Inc. † New York

CONTENTS

PREFACE

In 1874, Danish chemist, Christian Hansen, produced the first specimen of rennet by extracting dried calves' stomachs with saline solution which started the history of modern enzyme technology. In fact, this was the first enzyme preparation of relatively high purity used for industrial purposes. Ever since, the enzymes have come a long way in their evolution and have been used in in brewing processes, in baking, and in the production of alcohol. It is generally known that enzymes were already used in the production of cheese since old times. In 1876, William Kuhne proposed that the name 'enzyme' be used as the new term to denote phenomena previously known as 'unorganized ferments', that is, ferments isolated from the viable organisms in which they were formed. The word itself means 'in yeast' and is derived from the Greek 'en' meaning 'in', and 'zyme' meaning 'yeast' or 'leaven'. Nevertheless, the enzymes have spearheaded important research ventures and products ever since their use. Their most splendid use has been as fungal amylase in the 1950s, which was used in the manufacture of specific types of syrup, i.e., those containing a range of sugars, which could not be produced by conventional acid hydrolysis. Decades of research in biochemistry and biotechnology have boosted knowledge of enzymes for industries as well as research. Many new techniques have been established to modify enzymes or increase their yields. Novel techniques for purification of enzymes are constantly developing and so are being discovered new application of enzymes in medicine, food, research and multifarious applications in industries. This panoply of applications of enzymes has led to their important role now in the biotechnology industry.

The biotechnology industry has seen a leap forward from the traditional products to the waste based bioeconomy. Around the world, biotechnology industries are renewing traditional industries, from chemicals to forestry and automotive manufacturing with renewable products and processes. The technologies are now being tailored on patterns to offer environmentally cleaner solutions, using and reusing what is traditionally considered waste – corn stalks or even garbage - as feedstocks for consumer and industrial products, ranging from fuels to environmental cleanup solution.

Unprecedented increase in urban population worldwide has created demand for water, energy, food and mineral resources with concomitant generation of waste streams, greenhouse gases and the use of nutrients. The sustainable use of resources and water and waste management requires practical and reliable technologies, which can maximize the recovery of resources with minimal energy use and environmental impact. The use of enzymes is an important component of the industrial process with minimum ecological print.

This led to the third wave of biotechnology after the green and red wave, called the 'White Biotechnology". In fact, the impact of this biotechnology where enzymes play a key role has been confirmed by scientific studies and reports, such as the Organization for Economic Cooperation and Development (OECD's) report on the application of biotechnology to industrial sustainability and, most recently, by the World Wide Fund for Nature (WWF) report on the potential of industrial biotechnology to cut CO_2 emissions and help build a greener economy. Enzyme creation and optimization, to create faster and more effective processes, is a key area of industrial biotechnology.

Currently, enzyme technology has entered a phase in which new technologies, an increased understanding of fundamental biology, biochemistry and bioinformatics, are beginning to shape the discovery, development, purification, and application of biocatalysts to a much greater extent. This development is leading to new enzyme applications and increases the impact of enzyme technology in industry, more so in the value-addition of wastes which is fueling the bioeconomy. In fact, the enzyme technology is close to a major breakthrough, owing to many factors ranging from simple cost savings, the strongly increasing demand for chiral chemicals, the trend towards sustainable industrial development (less waste, less CO_2), and last, but not least, the opportunities created by emerging and advanced technologies.

This book covers different aspects of the enzymes in value-addition of wastes into various bioproducts, and raises question of the environmental and economical sustainability of the whole process. This will illustrate the importance of the enzymes in the sustainability loop of the resource reutilization and transformation to valuable bio-products. Enzyme technology is undergoing major gene reshuffling measures to create tailor made enzymes to meet specific needs and cut the primary substrate (example, lignocellulosic matter at a specific position). Enzymes, as used now are viewed as cost savers in the long run by creating less toxic and low concentration side-streams which had to be often disposed off and thus respecting the principles of a biorefinery.

Dr. Mausam Verma
CO_2 Solutions Inc.
Scientist
2300, rue Jean-Perrin
Québec, Canada G2C 1T9

Date: July 31, 2013

Satinder Kaur Brar
Associate Professor
Institut national de la recherche scientifique
Centre Eau, Terre & Environnement
Québec, Canada G1K 9A9

In: Enzymes in Value-Addition of Wastes
Editors: Satinder Kaur Brar and Mausam Verma

ISBN: 978-1-62948-968-1

Chapter 1

ENZYMES PRODUCTION DURING VALUE-ADDITION OF AGRO-INDUSTRIAL WASTES

Carlos Ricardo Soccol*[*], *Luciana Porto de Souza Vandenberghe, Michele Rigon Spier, Susan Grace Karp, Valesca Weingartner Montibeller, Denise Naomi Xavier Salmon, Francieli Goelzer, Michelle Cristina Tanoue Batista and Vanete Thomaz Soccol

Federal University of Paraná, Bioprocess Engineering and Biotechnology Department, Post-Graduation Program in Bioprocess Engineering and Biotechnology, Paraná, Brazil

ABSTRACT

Several agro-industrial wastes have been already applied for value-addition and costs reduction in the production of some important biomolecules, such as enzymes. Sugarcane bagasse, wheat and rice bran, soybean husks, coffee husks, cassava bagasse, citric pulp are examples of agro-industrial wastes that are used in the production of hydrolytic enzymes such as cellulolytic enzymes (cellulases, xylanases, mannanases, laccases), amylolytic enzymes (α-amylases and glucoamylases); phytases, pectinases or polygalacturonases, proteolytic enzymes (acid and alkaline proteases), and other enzymes using different fermentation techniques. Solid-state fermentation (SSF) and submerged fermentation (SFm) are the main techniques reported for enzymes production. Some important aspects of enzyme production and process development are described such as physical and chemical variables influencing enzyme synthesis. Besides, downstream operations for the separation, recovery and purification of these enzymes are also described for these different processes. Finally, the analysis and comparison of enzymes production by the three fermentation techniques using agro-industrial wastes are presented.

Keywords: Enzymes, agro-industrial wastes, solid-state fermentation, submerged fermentation

[*] E-mail address: soccol@ufpr.br. Phone number: +55 41 3361-3191 / fax: +55 41 3361-3695.

INTRODUCTION

Value-addition of agro-industrial wastes for enzymes production consists in the processing of residues, which may be used as substrates or media compositions for enzyme synthesis by different microbial genus. The use of agro-industrial wastes has economic and environmental benefits. Economic benefits are linked to the reduction of enzymes production costs due to the low commercial value and application of these residues. In some cases, the re-use of residues is also an alternative that contributes to environmental benefits by reducing their organic charge.

Considering that biomass is mainly composed of cellulose and hemicellulose, there are numerous possibilities for the employment of residues for enzymes production, for example, cellulolytic enzymes. These materials have all the needed nutrients for cell growth and enzyme production. Sometimes, it is necessary to add some supplements to the production medium [1, 2]. In some cases, substrates must pass through pretreatments (mechanical, chemical or biological) in order to be easily accessed by microorganisms [2].

In recent years, there has been an increasing trend towards efficient utilization and value-addition of agro-industrial residues, such as coffee pulp and husk, cassava husk, cassava bagasse, sugarcane bagasse, sugar beet pulp, apple pomace, declassified potatoes, etc [3]. Pandey et al. [2] reported that several bioprocesses have been developed that utilize these raw material for the production of bulk chemicals and value-added fine products such as ethanol, single-cell protein (SPC), mushrooms, enzymes, organic acids, amino acids, biologically active secondary metabolites, etc [3]. Not only the application of agro-industrial residues in bioprocess provides alternative substrates, but also helps solving pollution problems. Biotechnological processes, specially the solid-state fermentation (SSF) technique, have contributed for their utilization.

In this way, the next sections will provide some important information about the main commercial groups of hydrolyses, which can be produced with use of different agro-industrial wastes. It includes some examples of the employed fermentation techniques and process conditions developed by scientific researchers working in this area.

2. ENZYMES PRODUCTION USING AGRO-INDUSTRIAL RESIDUES

Different groups of hydrolytic enzymes are presented below highlighting their main action and specific catalysis reactions and the agro-industrial wastes that were lately employed for their production. Amylases, phytases, cellulases, xylanases, pectinases, mannanases, laccases, alpha-galactosidases will be then presented in order to give a brief idea of the important process and developments of enzymes production during value-addition of agro-industrial wastes.

2.1. Amylases

Amylases catalyze hydrolysis of starch molecules liberating diverse products, including dextrins and, progressively, small glucose polymers. They are divided in three groups: α-amylase, which cleaves the bonds in the interior of the substrate; β-amylase, which acts on the non-reducing extremities of the substrate; and amyloglucosidase, which liberates units of glucose from the non-reducing end of substrate molecules. This subitem focuses on microbial amylases, such as α-amylase and glucoamylases which may be produced using agroindustrial residues [4, 5].

2.1.1. Glucoamylases

Glucoamylases or amyloglucosidase [(1,4)-(1,6)-D-glucan glucohydrolase, E.C. 3.2.1.3] cleaves the 1,4 and 1,6 bonds of starch from the non-reducing end up to glucose [4, 5, 6]. As starch is an energy store in many vegetables occuring extensively in waste materials produced in large quantities in food processing plants, and causes pollution problems [7].

A SmF had been developed for the treatment of starch processing wastewater (SPW) with the production of fungal protein and glucoamylase enzyme. *Rhizopus oligosporus* DAR 2710 had the ability to convert more than 95% starch materials in SPW and achieved 9-10 $U.mL^{-1}$ in 14 h of cultivation at 35°C and initial pH 4.0 [7].

Three types of wastes generated during the processing of paddy (rice with the husks) to rice flakes at different stages of the manufacturing process have been categorized [8]. The first residue is the coarse waste, which is generated during shelling (after cleaning, soaking and roasting of paddy), contains husks and flattened broken pieces of rice. The second residue is the medium waste, generated during flaking and polishing, which generates remaining small particles of husk which get removed along with the rice bran and broken flakes as dust. The third residue is the fine waste, which is generated after drying and sieving gives fine husks and rice waste as dust.

Anto et al. [8] analyzed starch content of the raw materials and correlated it with the glucoamylase production under solid-state fermentation (SSF). Wheat bran (75.6% of starch content) and coarse waste (71.1% of starch content) presented highest glucoamylase production (264 $U.gds^{-1}$ and 211.5 U gds^{-1} respectively) compared to the enzyme production with rice powder (55.8%), medium waste (48.6%) and fine waste (34.2%). Higher enzyme production using wheat bran and coarse waste can be correlated with their starch content. The production media contained solid substrate and mineral solution (1:2 m/v), pH 7.0. Mineral solution contained $(NH_4)_2SO_4$, $MgSO_4.7H_2O$, $FeSO_4.7H_2O$, K_2HPO_4 and KH_2PO_4 inoculate with 1 mL (10^6 $spores.mL^{-1}$ collected from 72 h grown culture of *Aspergillus* sp. HA-2.

Food waste represents a considerable proportion of solid garbage discharged from several establishments such as restaurants, schools, cafeteries, and also from households. According to FAO [9], 1.3 billion of tons of food are discharged in all years, representing approximately one third of all food goes to waste. Industrialized and developing countries dissipate roughly the same quantities of food — respectively 670 and 630 million tones.

Wang et al. [10] used food waste (FW) for glucoamylase production by *A. niger* UV-60 under SFm. The residue was filtered out to remove excessive water. The fungus was cultivated for 4 days at 30ºC on the smashed-FW 2.50 (dry basis) and agar. Spores suspension 5% (v/v) (106 spores mL^{-1}) was used as inoculum. A SFm was carried out at 30ºC, natural pH, 120 rpm during 96 h achieving 137 $U.mL^{-1}$ of glucoamylase with raw food waste.

Tea waste, supplemented with minerals, was used for glucoamylase synthesis by *Aspergillus niger* NCIM 1248 under SSF. Maximum glucoamylase synthesis (198.4 $U.gds^{-1}$) was obtained on the mineralized tea waste at 96 h with an inoculum, initial moisture and pH of 4.0, 60% and 4.5, respectively at 30°C [11].

Ellaiah et al. [12] reported that glucoamylase yields with wheat bran were 247 U gds^{-1}, using fructose as additive 1% w/w, urea 1% w/w, at 30ºC during 120 h, inoculum level of 10% v/v, moisture content of 80%, 1:50 ratio of substrate weight to flask volume and pH 5.0.

2.1.2 α-Amylases

α-amylase (1,4-α-glucan-4-glucanohydrolase, EC 3.2.1.1.) breaks the α (1,4) bonds of polysaccharides that have or more units of D-glucose linked by α-1,4 bonds. The attack occurs in a non-selective form (as endoenzyme) on different points of the chain simultaneously, so that the first hydrolysis products are oligosaccharides of 5 to 7 units of glucose [4, 5].

Residues of wheat processing were used as substrate for amylase production. The medium was supplemented (in g.L^{-1}) of: peptone 24; urea 1.9; glycerol 1.5; KH_2PO_4 0.6 MgSO4 0.5; $(NH_4)_2SO_4$ 0.25 and distilled water. The initial pH was 5.0, and the medium was inoculated with 1.10^6 spores.mL^{-1} of a spore suspension from *A. oryzae* NRRL 695. SmF was carried out in a rotary shaker (150 rpm) at 30ºC for 96 h [13].

Ramachandran et al. [14] studied the production of α-amylases under solid-state fermentation (SSF) by *Aspergillus oryzae* using coconut oil cake (a by-product obtained after oil extraction from dried copra) as substrate. It contains starch, soluble sugars, soluble proteins, lipids and trace amounts of nitrogen. They achieved 3388 U.gds^{-1} when coconut oil cake was supplemented by 1% of peptone in 72 hours of fermentation at 30ºC, carried out with 2 mL spore suspension (6.10^8 spores.mL^{-1}) with the initial moisture content of 66%.

2.2. Phytases

Phytases (myo-inositol hexakisphosphate phosphohydrolase EC 3.1.3.8 and EC 3.1.3.26) belongs to a subclass of histidine acid phosphatases and catalyzes the hydrolytic degradation of myo-inositol hexakisphosphate (phytic acid) to free inorganic phosphate, to myo-inositol phosphate esters and free myo-inositol [15, 16]. This enzyme has been studied as a feed supplement for monogastric animals and foods for humans' consumption in order to increase the bioavailability of phosphorus and minerals. Also, phytases can decrease the excretion of phosphorus and, therefore, reducing environmental problems [17].

Soy whey was used in the medium composition for *Saccharomyces cerevisiae* MTCC 5421 fermentation to produce phytase, achieving 200 U.mL^{-1} after 36 hours of SFm, at 30ºC and pH 3.5.

Waste vinegar residue, the by-product of vinegar processing was used as substrate for phytase production from *Aspergillus ficuum* NTG-23 in SSF to investigate the potential of its re-utilization or recycling. Best phytase activity (98.2 U.g^{-1}) was attained using glucose (7.2%), soybean meal (5.1%), and incubation time of 271 h [18].

The utilization of crude glycerol, a by-product from biodiesel production, could bring significant economic and environmental benefits. It is the main by-product of the conversion of vegetable oils into biodiesel, comprising approximately 10% by mass of the vegetable oil fed to the system. Tang et al. [19] investigated a low-grade glycerol as the sole carbon source for phytase production with recombinant *Pichia pastoris*, achieving 60 U.mL^{-1}. Higher phytase activity level (1125 U.mL^{-1}) was achieved in 53 h of cultivation in fed-batch fermentation with crude biodiesel derived glycerol. Phytase from *A. oryzae* AK9 was produced by SSF using soybean meal as the main substrate at 30ºC for 96 h (15 $U.g^{-1}$) [20].

The wheat bran has been used as a low-cost phytate source and it has been frequently reported in the literature as a basal medium component for phytase production [21, 22, 17]. In this way, several studies were carried out. Salmon et al. [23] developed a bioprocess for phytase production by the basidiomycete *Schizophyllum commune* through SSF using wheat bran as substrate/support. After optimization studies, a maximal level of phytase (113.7 $U.g^{-1}$) was obtained in basal medium containing 5% sucrose, 50% moisture, 7.5% of biomass at 33ºC pH 7.0 during 72 h. Analysis of fermentation parameters profile for phytase production showed the highest productivity of 1.466 $U.gds^{-1}$ h^{-1} at 66 h of fermentation.

Bajaj & Wani [24] isolated the bacterium *Nocardia* sp. MB 36 from rice mill waste, which was producer of a thermostable and acid/alkali stable phytase. SmF was conducted at 47ºC under shaking of 120 rpm producing a maximum phytase after 40 h (0.254 $U.mL^{-1}$).

Phytase production from *A. ficuum* NRRL 3135 reached 15 $U.gds^{-1}$ with wheat bran by SSF. The addition of starch and $(NH4)_2SO_4$ increased the enzyme production to 25 $U.g^{-1}$ [25]. Chadha et al. [26] studied the optimization of phytase from *Rhizomucor pusillus* by SSF. The maximal level of phytase (9.18 $U.g^{-1}$) was achieved at 50 ºC, pH 6.0 and inoculation after 48 h of incubation with a 12 h old inoculum.

Gulati et al. [27] optimized the production of phytase by SSF using *Mucor indicus* MTCC 6333 achieving 266 to 513 $nkat.gdm^{-1}$. The thermotolerant fungus, *Aspergillus niger* NCIM 563, was also used for extracellular phytase production using wheat bran in SSF. Maximum enzyme activity was achieved (79.5 $U.g^{-1}$). The authors observed that the enzyme production was associated with high biomass growth and a corresponding reduction of phytic acid [28].

Phytase was produced using *Thermoascus aurantiacus* (TUB F 43) by SFm. A semisynthetic medium supplemented with 3.75% (w/v) wheat bran particles and 2% (v/v) Tween-20 was found to be the best production medium for maximum phytase activity (468.22 $U.mL^{-1}$) [22].

Aspergillus niger was investigated for phytase production in SmF and SSF. Overall phytase production increased with increasing shaker speeds from 150 to 300 rpm. Although specific growth rates and phytase production rates were higher at 150 rpm for 72 h from inoculation, increased productivities were obtained with inocula grown at 150 rpm and 300 rpm with gum addition [29].

A *Bacillus subtilis* US417 has been reported to produce an extracellular phytase activity (PHY US417) of 0.64 $U.mL^{-1}$ in SmF when grown for 20 h with the supplementation of wheat bran at 2% [30].

The phytase production by *Bacillus subtilis* US417 was carried out using wheat bran as substrate in SmF and SSF. The variables (inoculum size, methanol and yeast extract) were optimized and a high enzyme production of 112 $U.g^{-1}$ of wheat bran was attained. However, the phytase productivity was higher in SFm than in SSF [17].

Tahir et al. [31] reported that *Aspergillus niger* St-6 displayed maximum phytase activity (85 $U.mL^{-1}$) under SFm at 30ºC supplemented with 1% w/v of ammonium nitrate.

Shah et al. [15] studied phytase production from *Aspergillus niger* NCIM 563. The mutants N-1 and N-79 produced 7.619 and 9.523 $U.L^{-1}$ per day, respectively. Highest phytase activity of 80 $U.mL^{-1}$ was obtained with 1% rice bran; 3.5% glucose medium with 0.2 vvm aeration and 550 rpm at room temperature on the 8th day of fermentation. Citric pulp is an agro-industrial residue from the citrus processing industry applied in animal feed [32]. Spier et al. [33] evaluated the phytase production in SSF using citric pulp supplemented with a nitrogen source at 60% of moisture after 96 hours at 30°C. The highest phytase activity (51.53 $U.g^{-1}$) was achieved from the soil isolated *A. niger* FS3.

Neves et al. [34] studied a 2^3 factorial design using pH, temperature and moisture on the production of phytase by *Lichtheimia blakesleeana* URM 5604 through SSF of citrus pulp. Statistical analyses of the results showed that only the pH influenced the production of phytase, with the best production (264.68 $U.g^{-1}$) occurring at pH 6.0, 34°C, initial moisture 50%, after 48 hours of culture.

Phytase production was conducted in SSF by *Sporotrichum thermophile* TLR50 in sesame oil cake at 120 h and 45°C. Supplementation of sesame oil cake with glucose and ammonium sulfate further enhanced phytase activity (282 $U.g^{-1}$). Phytase production by *S. thermophile* was 22-fold higher in SSF (276 $U.g^{-1}$) than in SmF (12.60 $U.mL^{-1}$) [35].

The same authors investigated SSF using sesame oil cake as substrate for phytase production by *S. thermophile*. Glucose 3%, ammonium sulphate 0.5% and incubation period 120 h were identified as the most significant factors for the maximum phytase production. An overall 2.6-fold improvement in phytase production was achieved due to optimization. Higher enzyme production (348.76 $U.g^{-1}$) was attained on coated metallic trays [36]. Rodríguez-Fernandez et al. [37] evaluated phytase production by *Aspergillus niger* FS3 in SSF with citrus peel pilot scale under different aeration conditions. The best airflow intensity was 1 VkgM (L of air.kg of $medium^{-1}.min^{-1}$), which allowed the production of 65 $U.g^{-1}$.

Phytase production was studied using *Klebsiella* sp. DB-3 FJ711774.1 under SmF with orange peel as substrate. Orange peel flour and wheat bran (2% w/v) provided a maximal yield of phytase, 3.15 and 2.41 $U.mL^{-1}$, respectively, incubated at 50ºC for 72 h [38].

Production of phytase using canola meal by *Aspergillus ficuum* NRRL 3135 during SSF was studied. Optimum media moisture of 64% with inoculum age between 2 and 5 days enhanced the rate of enzyme production (4.71 $U.g^{-1}$) [21]. Another study [39] reported phytase production in rapeseed meal by *Aspergillus niger* during SSF for 120 h of incubation. By using 6% of glucose in SSF, the enzyme production reached 4.23 Ug^{-1}. The authors also studied the efficiency of gamma irradiation (1 kGy) on phytase production which attained 6.5 $U.g^{-1}$. Castor beans residue was used to produce phytase by *Paecilomyces variotii*. Maximum phytase activity was 260 $U.g^{-1}$ after 72 h [40].

The production of phytase by *Pseudomonas* AP-MSU 2 was also investigated in SSF carried out at 32°C for 72 h, 5 g of cassava waste meal, 70% moisture and inoculum size of 2 mL. Phytase production reached 32 $U.g^{-1}$ [41].

Cassava dregs were employed for phytase synthesis from *A. niger* PD. Ammonium nitrate was the best nitrogen source. Sodium dodecyl sulfate promoted phytase yield of 6.73 U g^{-1} of dry mass [42]. Dry olive waste, an agro-industrial material derived from the two-phase-extraction of olive oil was used as a substrate for phytase production by *Aspergillus niger*. SSF was performed at 30ºC for 120 h. In the optimized process conditions (moisture 70%; corn steep liquor as a nitrogen source; inoculum size of 3–4 mL; presence of phosphate), *A. niger* was able to produce 58 $U.g^{-1}$ [43].

Roopesh et al. [44] used SSF for phytase production using various substrate combinations of wheat bran and oil cakes by *Mucor racemosus* in SSF. The combination of wheat bran and oil cake yielded larger phytase production (44.5 $U.g^{-1}$).

A bacterial strain, isolated from soil, designated as PH01 was capable of producing a thermo-acido-tolerant phytase. A complex medium for PH01 phytase production was was composed by rice bran and soybean meal extract containing 3% (v/v) phosphate solution (7.3% $NaHPO_4$ + 3.2% KH_2PO_4, pH 7.2). Fermentation was carried out at 37ºC with aeration for 48 h achieving 10 $U.mL^{-1}$ [45].

The linseed oil cake and wheat bran (1:1) were used for phytase production by *Rhizopus oryzae*. This mixing showed the highest phytase yield (17.68 $U.g^{-1}$) at 96 h in SSF. After optimized studies, an overall 8.41-fold increase in phytase production was achieved at the optimum concentrations (mannitol, 2.05%; ammonium sulfate, 2.84% and phosphate, 0.38%, w/w) and a reduction in fermentation time from 96 to 72 h [46].

Bhargav et al. [47] and Ramachandran et al. [48] reported that cottonseed oil cake and olive oil cake poorly supported phytase production individually. However, mixed substrate fermentation increased enzyme production to 35 $U.g^{-1}$. Supplementation of glucose further enhanced phytase activity to 52 $U.g^{-1}$.

2.3. Cellulases

Cellulase refers to *enzymes* which catalyzes cellulose hydrolysis. There are five general types of cellulases based on the type of reaction they catalyze: endocellulases (E.C. 3.2.1.4), exocellulases (E.C. 3.2.1.91), cellobiase (E.C. 3.2.1.21), oxidative cellulases and cellulose phosphorylases.

Brazil is the largest producer of sugarcane with 493 million tons [49]. The Centre-South region of Brazil accounts for almost 80% of feedstock production. The Brazilian bioethanol industry was poised for a major jump during 2006–2008 as a part of new national plan to increase the sugarcane production by 40% by 2009 [50]. Sugarcane bagasse is a residue of cane stalks left over after the crushing and extraction of the juice from the sugarcane [2]. It presents a great morphological heterogeneity and consists of fiber bundles and other structural elements such as vessels, parenchyma, and epithelial cells [51]. It is composed of 19-24% of lignin, 27-32% of hemicellulose, 32-44% of cellulose and 4.5-9.0% of ashes. The remainder is mostly lignin with lower amounts of minerals, waxes, and other compounds [52]. Sugar mills generate approximately 270-280 kg of bagasse (50% moisture) per metric ton of sugarcane [53]. The Brazilian annual production of sugarcane bagasse is currently estimated at 186 million tons of bagasse [54].

The conversion of sugarcane bagasse into fermentable sugars requires essentially two steps: pretreatment and hydrolysis. Pretreatment is usually performed by thermo-chemical processes. Although the pretreatment is required to make the biomass accessible to the enzymes action, it is desirable to use mild conditions that minimize the degradation of the sugars and lignin into inhibitory by-products [55]. Therefore, to improve the enzymatic hydrolysis process and offset the low severity applied during the pretreatment, the trend is to use enzyme mixtures containing cellulases, xylanases and other accessory enzymes [56]. White-rot basidiomycetes produce several ligninolytic enzymes (laccases, manganese peroxidases and lignin peroxidases) that catalyze one-electron oxidation of lignin units, producing aromatic radicals and promoting lignin degradation [57].

The production of cellulases using sugarcane bagasse and soybean meal was reported. Cellulose activities in CMCase and FPase reached 379 $U.gds^{-1}$ and 197 $U.gds^{-1}$, respectively [58].

Sugarcane bagasse is being used by many authors in the production of enzymes [59, 60, 61, 62]. Membrillo et al. [63] cultivated *Pleurotus ostreatus* using not pretreated sugarcane for extracellular cellulases production, CMCase (0.18 $U.gds^{-1}$) and PFase (0,018 $U.gds^{-1}$). The fibrous nature of sugarcane naturally favors the induction of enzymes of cellulolytic complex.

2.4. Xylanases

Xylanases are glycosidases (*O*-glycoside hydrolases, EC 3.2.1.x) which catalyze the hydrolysis of 1,4-D-xylosidic linkages in xylan [64]. A complete cleavage of glycoside linkages in the heteroxylan backbone requires the interaction of a number of the main-chain- and side-chain-cleaving enzyme activities. Endo-1,4-xylanase (1,4-D-xylan xylanohydrolase, EC 3.2.1.8) and xylan-1,4-xylosidase (1,4-D-xylan xylohydrolase, EC 3.2.1.37), or exoxylanase, are enzymes capable of cleaving the main-chain glycosyl groups [65, 66].

Corn steep liquor (CSL) is a by-product of corn wet milling process. CSL is a mixture of soluble proteins, free amino acids, reducing sugars, organic acids (e.g., lactic acid), vitamins, and minerals [67], and its composition depends on the corn variety and steeping conditions.

Fang et al. [68] used corn steep liquor for increasing a low molecular weight xylanase production in SSF of by *Aspergillus carneus* M34. After optimization study, the optimal media composition was: coba husk/corn steep liquor (4.5/0.5) in $g.L^{-1}$: NH_4NO_3 32.4; $CaCl_2$ 1.34; $MnSO_4.7H_2O$ 0.0124. The optimized xylanase activity of 1721 $U.g^{-1}$ was obtained on the 6^{th}day at 30°C, initial moisture content of 75%, which was an increase of 227% in xylanase production. The incubation time was reduced from 12 to 6 days by increasing inoculums size from 2.10^6 to 2.10^7 $spores.mL^{-1}$.

Studying the effect of orange pomace, orange peel, lemon pomace, lemon peel, apple pomace, pear peel, banana peel, melon peel and hazelnut shell on the production of xylanase, using *Trichoderma harzianum* 1073-D3, Seyis & Aksoz [69] discovered that molasses are able to reduce the time of production in 50% when used as an additional carbon source. The maximum activity has been observed on 2.5% melon peel medium (26.5 $U.mg^{-1}$ of protein) incubated at 30°C for 7 days on a rotary shaker (150 rpm).

Rose & Zyl [70] optimized xylanase production using a recombinant strain of *A. niger* D15[xyn2]pyrG$^-$. The highest xylanase activities of 226 and 209 $U.mL^{-1}$ were produced with 20 and 30% molasses, respectively, at 30ºC, pH 6.5, agitation of 100 rpm and a spore inoculum of 1.10^6 $spores.mL^{-1}$.

The sugarcane bagasse was tested as low cost raw material for xylanase production. They used *Bacillus circulans* D1 under SFm containing mineral medium and hydrolysated of bagasse with initial sugar concentration of 2.5 $g.L^{-1}$, agitation of 200 rpm, incubated at 45°C. The authors achieved 8.4 $U.mL^{-1}$of xylanase in 24h of cultivation [71]. Meshram et al. [72] produced xylanase by SFm using *Penicillium janthinellum* NCIM in 50 mL of Mandels–Weber medium, sugarcane bagasse (particle size of 40#) and beef extract, incubated at 28–30°C at 180 rpm. The best parameters observed were: carbon source 1.63%, nitrogen source 0.16%, pH 4.1, and inoculum 5.5%, achieving a maximum xylanase activity of 28.98 $U.mL^{-1}$.

Maciel et al. [66] also produced xylanase with sugarcane bagasse under SSF by *Aspergillus niger* LPB 326. The highest xylanase activity was 2,327 U.gdm^{-1} using 65% of sugarcane bagasse and 35% of soybean meal supplemented with a mineral salt solution, 85% initial moisture, 10^{6} spores.gdm^{-1},at 30°C for 4 days.

Paddy husk was used as support to optimize xylanase production by SSF using *Bacillus pumilus*. The medium contained 200 g of paddy husk with 800 mL of liquid fermentation medium (20 g.L^{-1} of xylan; 2 g.L^{-1} of peptone; 2.5 g.L^{-1} of yeast extract and mineral solution at pH 9.0). The highest xylanase activity of 142 U.gdm^{-1} was obtained after 6 days of fermentation at 30°C. The xylanase activity was highest (147.3 U.gdm^{-1}) with a 2:9 ratio of paddy husk and liquid fermentation medium on the 6th day at 40°C achieving 177.5 U.gdm^{-1} [73].

Expressive amounts (more than two million tons per year) of agricultural wastes are generated from the coffee processing industries, such as coffee pulp, coffee cherry husk, silver skin and spent wastes [74]. These substrates contain carbohydrates, proteins, fibers and minerals with potential for utilization in bioprocesses [2].

Production of xylanase utilizing coffee by-products was reported for the first time by [74]. They used coffee pulp, coffee cherry husk, silver skin and spent coffee as a sole carbon source for the production of xylanase by SSF using *Penicillium* sp. CFR 303. Coffee cherry husk was the best substrate for xylanase production in 4 days. After optimization of fermentation parameters, the enzyme activity reached 19,560–20,388 U.g^{-1} with 50% of moisture, pH 5.0, temperature 30°C, particle size of 1.5 mm, inoculum size 20%, 5 days, supplemented with xylose and peptone. The enzyme production was further improved to 23,494 U.g^{-1} after pretreatment with steam at 121°C during 30 min.

According to AACC [75], barley bran is a product of roller or separation milling of dehulled or hulless barley, which includes the outer covering such as pericarp, testa (seed coat), aleurone and subaleurone layers. It provides around 5.5% (dwb) of soluble fiber and 15% of a total dietary fiber. Should be free of hulls and may also be a combination of milled bran and short fractions.

Haddar et al. [76] tested barley bran as substrate for xylanase production by *Bacillus mojavensis* A21. After optimization, 7.45 U.mL^{-1} of xylanase were obtained with 18.66 g.L^{-1} of barley bran, NaCl 1.04 g.L^{-1}, 176 rpm and 34.08 h of cultivation at 50ºC and pH 8.0.

Xu et al. [77] produced xylanase with a mutant *Pseudomonas* sp. WLUN024 grown on xylosidic materials, such as hemicellulose, xylan, xylose, and wheat bran. Batch fermentations were carried out on a rotary shaker at 220 rpm, 37°C for 24 h. After optimization the maximum activity of xylanase reached 1245 U.mL^{-1}. The optimized medium consisted of 70 g wheat bran, 8 g $(NH_4)_2$ SO_4 and 4 g.L^{-1} K_2HPO_4 and initial pH adjusted to 8.5 before autoclaving.

Michelin et al. [78] studied cultivation of *Aspergillus terricola* in stirred tank and airlift bioreactors using wheat bran particles suspended in the cultivation medium as substrate for xylanase and β-xylosidase production. The conditions used on stirred tank bioreactor were 30°C, 300 rpm, aeration of 1 L.min^{-1} reached 7,475 U.L^{-1} of xylanase activity after 36 h of operation. In the airlift bioreactor, xylanase production was higher, achieving 12,845 U.L^{-1} with 1 L.min^{-1} of aeration rate.

Another study with wheat bran as a substrate for xylanase production was conducted by Kapoor et al. [79]. *Bacillus pumilus* SV-85S was grown in SSF. High level production (73,000 U.gds^{-1}) of an extracellular cellulase-poor alkali stable xylanase has been obtained after 48 h at 30°C with 1:3 ratios of substrate to moistening agent using wheat bran as a carbon source. The enzyme was completely stable over a broad pH (5–11) range and retained 52% of its activity at a temperature of 70°C for 30 min.

Recent studies on the feasibility of using citrus peel for producing enzymes were carried out [80]. The authors produced xylanase in SSF process by *Aspergillus niger* F3 in a 2 kg bioreactor testing different aeration conditions to optimize production. The air flow intensity of 1 V kg M (volumetric airflow per kilogram of medium) allowed a production of 65 $U.g^{-1}$ xylanase with 60% of initial moisture, initial pH 5.0 and 30°C after 96h.

Diaz et al. [81] evaluated the potential of washed grape pomace and orange peels to be used as substrate in SSF and SmF for the production of several hydrolytic enzymes. Increased xylanase synthesis was obtained with the addition of orange peels in a 1:1 proportion with grape pomace. SSF was carried out in static conditions at 30°C, initial moisture content of 70%, inoculum concentration of 4.5×10^{8} $spores.gdm^{-1}$ using 10 g of the sterilized solid substrate. SmF was carried out for 8 days in a 5 L vessel (2.5 L work volume), at 30°C, pH maintained at 3 by the addition of 1 M HCl, air flow rate of 2.5 $L.min^{-1}$ and agitation of 120 rpm. In SmF, xylanase production reached a maximum of 98±14 $U.g^{-1}$ at the fifth day.

The apple pomace is a waste resulting of apple agro industry [82]. The apple pomace composition on dry weight basis is 3.97-5.40% moisture, 2.54-3.28% malic acid, 48.00-62.00% total carbohydrate, 8.53-18.50% vitamin C, 3.29% soluble proteins, 4.45-5.67% protein, 4.70-48.72% crude fiber, 3.49-3.90% fat and 3.50-14.32% pectin [83]. Considered as a promising substrate for culturing microorganisms in SSF, apple pomace (AP) has been already used for lignocellulolytic enzymes, such as cellulases and pectinases.

Using apple pomace as a substrate in SSF, the effect of different inducers on cellulase and hemicellulase bioproduction by *Aspergillus niger* NRRL 567 were investigated [84]. Apple pomace was supplemented with 1% (w/w) rice husk. The substrate was inoculated with spore suspension of 1×10^{7} spores/g substrate and inducers were added in different concentrations. The initial moisture was adjusted to 75% and cultures were incubated at 30±1°C for 7 days. A rapid production of xylanase and different cellulose enzymes - FPase , CMCase and β-glucosidase were observed, with peak activity reaching between 48 and 72 h of fermentation. The higher xylanase activity of 1412.58 $U.gds^{-1}$ was observed with veratryl alcohol (3 $mM.kg^{-1}$) followed by 1247.84 $U.gds^{-1}$ with $CuSO_4$ (1 $mM.kg^{-1}$) inducer after 72 h of fermentation.

In another study in the same year, Dhillon et al. [84] using apple pomace (AP) as solid substrate by *Aspergillus niger* NRRL 567 and *A. niger* NRRL 2001 tested the potential of lactoserum as a crude inducer and moistening medium for fungal cellulase and hemicellulase induction through SSF. Xylanase activity of 4868 $U.gdm^{-1}$ was obtained using *A. niger* NRRL 567 and 4580.7 $U.g^{-1}$ xylanase activity for *A. niger* NRRL 2001 carried out at 30°C for 5 days. For enzyme production, the AP with 70% (v/w) initial moisture was prepared by moistening with lactoserum (LAC) at different ratios. Higher enzyme activities were achieved after 96h, AP treatments (LAC:H_2O) ratio 5:5, and with 50 $mol.m^{-3}$ citrate buffer (pH 4.8, 15:1, v/w) for enzyme extraction. Best results were obtained when AP was supplemented with LAC.

Banana peel has high energy content but is low in protein and amino acid. It has about 7-12% cellulose, 6.4-9.6% lignin, and 6.4-8.4% hemicelluloses (Wachirasiri, 2009), 0.90% protein, 1.7% crude lipid, 59.0% carbohydrate and 31.70% crude fibre (Anhwange, 2008).

In 2011, Yang and collaborators [85] utilized banana peels in SmF with multi-funcional cellulase activities genes expressed in *Coprinopsis cinerea*. The fermentation of lignocellulosic substrates was carried out using mineral solution along with dry banana peels at a concentration of 40 g.L^{-1}, 150 rpm and 27°C and initial pH of 6.0. The highest activities of endo-b-1,4-glucanase (0.48 U.mL^{-1}), total cellulase (0.26 U.mL^{-1}) and endo b-1,4-xylanase (38.10 U.mL^{-1}) were observed after 7 days. Enzymatic activities were increased nearly two fold in comparison with the wild-type strain.

Brewers' spent grain (BSG) is a by-product of mashing process. BSG is the residue obtained after the wort has been removed by filtration and it is mainly used as low value cattle food [86]. The chemical composition of BSG produced from 100% malted barley is hemicelluloses (284 g.kg^{-1}), lignin (278 g.kg^{-1}), cellulose (168 g.kg^{-1}), proteins (152.5 g.kg^{-1}), ashes (46 g.kg^{-1}) and extractives (58 g.kg^{-1}). The hemicellulosic structure is formed by the xylan and arabinan polysaccharides in the proportion 2.3:1 [87].

Looking for agro-industrial by-products capable of producing industrially valuable enzymes, Mandalari et al. [86] studied brewers' spent grain (BSG). *Humicola grisea* var. *thermoidea* and *Talaromyces stipitatus* were studied for xylanase production. The maximum activities of xylanase (16.9 U.mL^{-1}) from cultures of *H. grisea* were achieved at 150 rpm, 45°C and 8 days on BSG. A maximum feruloyl esterase (FAE) activity of 0.47 U.mL^{-1} on BSG were also identified in the same conditions. After microbial growth an analysis of residual cell wall material was made and showed the solubilisation of arabinoxylan and cellulose, two main polysaccharides present in BSG, showing that BSG could be used as source of xylanase.

Oliveira et al. [88] tested xylanolytic enzymes production by *Penicillium janthinellum* with five agricultural wastes in SFm. Among the wastes, corn cob was the best for xylanase production (55.3 U.mL^{-1}) in shake flask 60 rpm, 30°C during 132 h, initial pH 5.5 (uncontrolled). This level of xylanase activity is comparable or higher than that reported for most *Penicillium* in SmF.

Dobrev et al. [89] also used corn cobs in medium composition for increasing xylanase production by *Aspergillus niger* B03. The optimization process was performed. The fermentation was carrying out in flasks inoculated with 10% inoculums, cultivated at 28°C for 64 h at 180 rpm shaking. The xylanase activity obtained with the basic nutrient medium was 750.37 U.mL^{-1}. After optimization, xylanase activity attained 996.3 U.mL^{-1}, which was 33% higher than the activity obtained with the basic medium. The nutrient medium optimized was composed (in g.L^{-1}) by $(NH_4)_2HPO_4$ 2.6, urea 0.9, corn cobs 24, wheat bran 14.6 and malt sprout 6.

Searching for improvement of xylanase production, Su et al. [90] utilized corn cobs and the thermophilic fungus *Thermomyces lanuginosus* SDYKY-1. The medium composition was optimized for improving the thermostable xylanase using economical carbon and nitrogen sources. The authors evaluated the effects of nine variables (powdered corncobs, soybean meal, Tween-80, $CaCl_2$, $MgSO_4.7H_2O$, $FeSO_4$, KH_2PO_4, initial pH and inoculum culture volume). The experiments proved that corn cobs, soybean meal and $FeSO_4$ significantly influenced xylanase production, and then these three factors were therefore optimized. The fermentation flasks were incubated at 50°C at 210 rpm for 72 hours and a maximum xylanase production of 3078 U.mL^{-1} was obtained with concentration of corncobs 38.7 g.L^{-1}, soybean meal 17.5 g.L^{-1} and $FeSO_4$ 0.26 g.L^{-1}, representing a 144% increase on xylanase production.

The higher xylanase activity and the lower production cost may provide corn cobs a further application in the industrial scale.

Heck et al. [91] determined the optimal conditions of cultivation time, aeration and temperature for xylanase production by *Bacillus circulans* BL53 with industrial fibrous soy residue (IFSR). Bioreactor experiments were carried out in a 500 mL cylindrical bioreactor loaded with 20 g of dry IFSR soaked in 100 mL of basic liquid medium and inoculated with 20 mL of inoculum which had its OD adjusted to 1.0. The results after the application of a 2^3 central composite design (CCD) suggest that xylanase production by *B. coagulans* BL53 is higher at moderate cultivation aeration (500 $mL.min^{-1}$, which corresponds to 25 VMM) and at low temperatures (25ºC) by 5 days. The predicted optimal parameters were tested in the laboratory and the final xylanase activity obtained was very close to the predicted value (0.928 $U.mg^{-1}$, predicted; 0.953 $U.mg^{-1}$, tested). Following CCD modeled conditions, it was able to increase 2.5-fold enzyme activity previously obtained (0.437 Umg^{-1}).

Antoine et al. [92] tested five agro-industrial wastes (soya oil cake, soya meal, wheat bran, whole wheat bran, and pulp beet for xylanase production. The fungus, *Penicillium canescens* was investigated in SSF. A xylanase production level of 18,895 $U.g^{-1}$ in Erlenmeyer flasks and 9,300 $U.g^{-1}$ in plastic bags were reached after 7 days of incubation with 5g of soya oil cake crushed (5 mm particle size) supplemented with 20 mL of distilled water, 3% of casein peptone, 4% $Na_2HPO_4.2H_2O$ at 30 °C and 80% of initial moisture.

2.5. Pectinases

Pectinase (EC 3.3.1.15) is a group of enzymes which can hydrolyze the pectic bonds found in the cell wall of plants. It is widely employed in various industries.

Studies on pectinase production in stainless steel trays reported by Couri et al. [93] selected the wheat bran as the best substrate among the substrates tested (banana peel, mango peel, cassava bran and wheat bran) for maximum polygalacturonase production (30.75 $U.mL^{-1}$) at 42 hours of fermentation while mango peel had an opposite influence. The trays contained 40 g of sterile medium inoculated with 10^5 $conidia.g^{-1}$ incubated at 32°C.

Pectic enzymes are used industrially in fruit processing to facilitate pressing and to aid separation of the flocculent precipitate by sedimentation, filtration or centrifugation in the extraction of the clarified juice [94].

Rodríguez-Fernández et al. [80] used citrus peel (citric pulp) for pectinase (265 $U.g^{-1}$) production in SSF by *Aspergillus niger* F3 [95] in a 2 kg bioreactor with initial humidity of 60%, pH 5.0, air flow 1 V kg M (volumetric air flow per kilogram of medium).

Pectinases production by *Thermoascus aurantiacus* in SSF was reported by Martins et al. [96] using wheat bran. The authors reached 43 $U.g^{-1}$ of polygalacturonase at the 4th day of fermentation while pectin lyase (11600 $U.g^{-1}$) was produced in the 14th day. The same authors reported the production using orange bagasse (composed in dry material 11.8% fibre, 6.4% protein, 63% nitrogen, 6.7% ash, 19% total sugar (9% reducing sugar) and 0.1% pectin), reaching 43 $U.g^{-1}$ of polygalacturonase after 6 days and 19320 $U.g^{-1}$ of pectin lyase. SSF was conducted with initial moisture of 67%.

Patil & Dayanand [97] reported pectinase production using deseeded sunflower head under SFm by *A. niger* DMF 27 and *A. niger* DMF 45 in SSF. In both fermentations processes the pH of 5.0 and temperature of 34ºC were ideal. Under optimum conditions, maximum production of exo-pectinase was 34.2 $U.g^{-1}$ in SSF (65% initial moisture) and endo-pectinase was 12.6 $U.mL^{-1}$ in SmF.

2.6. Mannanases

Sugar beet molasses is the by-product obtained after the final crystallization process in sugar industry. It is used in the production of valuable industrial products such as ethanol, and microbial enzymes sources including fungi [98].

The optimal conditions for enhanced β-mannanase production by recombinant *Aspergillus sojae* were studied by Ozturk et al. [99]. Different concentrations of sugar beet molasses were used as substrate under submerged fermentation in shaken flasks. β-mannanase activity (363 $U.mL^{-1}$) reached its maximum at the 4th day at 30°C, 207 rpm with medium composed by (w/v): 7% sugar beet molasses, 0.43% NH_4NO_3, 0.1% K_2HPO_4 and 0.05% $MgSO_4$.

The holocellulose portion (combination of cellulose and hemicelluloses) at close associations with lignin, form a material with complex lattice structure described as lignocellulose. As the lignin portion forms a protective barrier around the holocellulose components of the plant biomass, consequently, cellulose and hemicellulose polymers are protected from enzymatic and microbial degradation [100].

Beukes & Pletschke [101] studied the synergistic action of sugarcane bagasse (SCB) pre-treated with alkaline salts with different ratio of enzymes for efficient hemicelluloses hydrolysis. The greatest amount of reducing sugar (1194.88 $\mu mol.min^{-1}$) and largest degree of synergy (2.85) was obtained using NH_4OH pre-treated SCB with the combination of two recombinant enzymes (25% ManA and 75% XynA) expressed in *Escherichia coli* cells.

Soybean (*Glycine max* (L.) Merrill) is one of the most important agricultural products in the world, reaching the production of 263.7 million tons in 2010/2011 [102]. Soybean meal is widely used in animal nutrition and as a supplement [103] generated in the mechanical and/or chemical oil extraction process. The protein content of soybean bran can vary between 42.5 to 55.6% depending on the employed oil extraction process.

Heck et al. [91] optimized the mannanase production by *Bacillus circulans* in SSF using industrial fibrous soy residue and nutritive solution in a 500 mL cylindrical bioreactor achieving 0.54 $U.mg^{-1}$.

Vandenberghe et al. [104] deposited in Brazil a patent that describe the mannanase production using soybean husks and *Aspergillus niger* non mutant strain. The authors achieved a mannanase activity of 200 $U.gdm^{-1}$ and 3600 $U.L^{-1}$ at the SSF and semi-solid state fermentation, respectively.

Valášková & Baldrian [105] cultivated *Pleurotus ostreatus*, *Trametes versicolor* and *Piptoporus betulinus* in SSF using wheat straw for mannanase production, reaching 35 $mU.gdm^{-1}$.

A mutant strain *A. niger* E-30 with high- and stable-yield β-mannanase was obtained from the parent strain *A. niger* LW-1. High levels of activity were reaching 36.675 $U.g^{-1}$ in SSF using wheat straw and a nutritive solution [106].

Ferreira & Filho [107] also observed the mannanase production by *Trichoderma harzianum* with wheat straw, and the maximum activity (9 $U.mL^{-1}$) was obtained between the 8th and 9th day.

The global increase demand for vegetable oils is responsible by the tremendous increase in palm oil production at last decade. Malaysia, Indonesia and Brazil are the three largest producers of palm oil [108]. Malaysia produces 41% of oil palm in the world, and its growth is reducing in relation to Indonesia, which corresponds to 43% and achieved the highest world producer in 2006 [109].

Mannose is the principal carbohydrate present in palm oil kernels with content in the range of 30–35%, but also 7–9% glucose is present and the extraction process of oil from palm kernels generate a residue known as palm kernel press cake (PKC). This process can be either mechanical or solvent extraction resulting in a residue containing around 50% carbohydrate and 15–20% protein [110, 111].

Heidorne et al. [112] studied the cellulases and hemicellulases produced by *Ceriporiopsis subvermispora* cultured on wood chips of *Eucalyptus grandis* and *Pinus taeda* by SSF. *Pinus taeda* showed better results with a specific mannanase activity of 90.4 $U.kg^{-1}$ at 13th day. Magalhães & Milagres [113] used a combination of ion exchange chromatography and SDS-PAGE data, and revealed that this mannanase had a high-molecular-weight of 150 kDa, active against galactoglucomannan and xylan.

Mannanase production was carried out with *Aspergillus siydowii* grown on 1.0% (w/v) banana stem as the carbon source, producing 1.229 $U.mL^{-1}$, and *Emericella nidulans* grown on 1.0% (w/v) dirty cotton residue producing 0.455 $U.mL^{-1}$ of mannanase activity [114]. Fermentation was conducted using as inoculum 10^8 $spores.mL^{-1}$ during 7 days and agitation of 120 rpm.

Xu et al. [115] studied the lignocellulose degradation and enzyme production by *Irpex lacteus* CD2 during SSF of corn stover (CS). The moisture of 0.425–0.85 mm pieces of CS was adjusted to 75% with water, and the ratio of CS to culture was 1:1 (w/v), incubated at 28°C. Mannanase reached its maximum activity level of 3.1 $U.g^{-1}$ of culture after 60 days.

Using defatted copra and synthetic medium, Lin & Chen [116] studied mannanase production by *Aspergillus niger*, under agitation. The maximum activity of this enzyme was 28 $U.mL^{-1}$.

The patent CN1266096 relates a mannanase activity of 500 $U.mL^{-1}$ produced by basiphilic *Bacillus* using konjac and soybean flour as raw materials in the fermentative process [117].

More recently, the patent CN101481674 relates the mannanase production by *Aspergillus niger* in solid medium containing wheat bran, soybean meal and konjac powder at a weight ratio 60–90: 10–40: 1–6. The final enzyme activity was $1.3x10^5U.g^{-1}$ for use in feedstuff [118].

2.7. Laccases

Laccases are multi-copper containing oxidases (EC 1.10.3.2), which catalyses the oxidation of phenols, polyphenols and anilines with the concomitant reduction of oxygen to water in a four-electron transfer process. In the presence of small redox mediators, laccase offers a broader repertory of oxidations including non-phenolic substrates. Hence, fungal laccases are considered as ideal green catalysts of great biotechnological impact due to their few requirements and their broad substrate specificity, including direct bioelectrocatalysis [119].

The production of laccases from molasses wastewater or vinasse was reported [120]. Two fungi, *Coriolus versicolor* and *Funlia trogii*, were cultivated on filtered and sterilized vinasse diluted 1:5. The maximum laccase activity of 4.880 $U.mL^{-1}$ was produced by *F. trogii* after 16 days in the presence of cotton stalk (1g per 30 mL). For the fungus *C. versicolor*, the highest enzyme activity of 4.406 $U.mL^{-1}$ was obtained in the same conditions.

Ntougias et al. [121] studied the biodegradation of olive mill wastewater with concomitant production of laccases and manganese independent peroxidases by 4 strains of basidiomycetes: *Ganoderma australe, Ganoderma carnosum, Pleurotus eryngii* and *Pleurotus ostreatus*. Fungal growth resulted in reduction of phenolic compounds by 64-67% and 74-81% for *Ganoderma* and *Pleurotus* spp., respectively. Maximum laccase activity of 117.82 $U.L^{-1}$ was produced by *P. ostreatus* after 30 days. *P. eryngii* produced 73.95 $U.L^{-1}$ after 20 and 106.85 $U.L^{-1}$ after 30 days of incubation. Lakthar et al. [122] evaluated the growth of *Lentinula edodes* on olive mill wastewater for polyphenol biodegradation and reached after 15 days (1.6 $U.mL^{-1}$) of laccase production.

Sugarcane bagasse is one of the most promising substrates for the production of laccases by SSF, for presenting significant lignin content (around 20%), being porous and presenting an absorption capacity of at least 1:10 (w/v). Laccases were produced by SSF of sugarcane bagasse by the selected strain *Pleurotus ostreatus* (coded Pl 22 Em) [123]. The laccase activity produced by *P. ostreatus* 22 in SSF was significantly affected by the concentrations of nitrogen source and two inducers of laccase expression, copper sulfate and ferulic acid. The maximum laccase activity (161.3 $U.g^{-1}$) would be obtained at the following conditions: yeast extract 6.417 $g.L^{-1}$, $CuSO_4$ 172.6 mM and ferulic acid 1.860 mM. Experimentally, the maximum laccase activity of 151.6 $U.g^{-1}$ was produced, under optimized conditions, at the 5th day of fermentation. Dwivedi et al. [124] produced laccase and xylanase on sugarcane bagasse through the co-cultivation of *Penicillium oxalicum* and *Pleurotus ostreatus*. A combination of sugarcane bagasse and black gram husk in a ratio of 3:1 was the best solid substrate and support for enzyme production. Maximum levels of xylanase (8205.31 $U.g^{-1}$) and laccase (375.53 $U.g^{-1}$) after 12 days of SSF were obtained by using 4 g of solid support with 80% moisture.

Sen et al. [125] demonstrated that the addition of sugarcane bagasse powder (1% w/v) enhanced laccase activity in a *P. ostreatus* liquid culture by 6-fold on the 7th day. The addition of bagasse alcoholic extract resulted in a 12-fold increase in laccase activity in the liquid culture and a 32-fold increase in the semi-solid culture. Moreover, sugarcane bagasse extract enhanced the storage life of laccase.

Some authors have reported the production of cellulases with the use of low cost substrates: [63, 105, 126, 127, 128, 129, 130] and others. These authors have used agro-industrial residues for celullase production using fungal strains and basidiomycetes of the genus *Pleurotus* and *Lentinus* attaining interesting cellulase activities (Table 1).

2.8. Alpha-Galactosidases

The α-D-galactoside galactohydrolase (EC 3.2.1.22), also named α-galactosidase, catalyzes the hydrolysis of a-1,6- galactosidic bonds, releasing a-D-galactose. These bonds are found in oligosaccharides such as melibiose, raffinose and stachyose, which are not digerible by humans and monogastric animals. These enzymes may be applied in food and feed processing to improve digestibility to human and animal nutrition [136, 137].

Vinasse is the waste-product of ethanol recovery by distillation. Ethanol is produced by the alcoholic fermentation of soybean molasses, an industrial residue, produced concomitantly with the soybean protein-concentrate, and results from the extraction ofsugars from de-oiled soybean meal. The composition of soybean vinasse is (in average): stachyose 11%, raffinose 22%, galactose 1.8%, proteins 13.3, lipids 27.8, ashes 9.2, fibers 14.6 [137] .

Sanada et al. [137] studied the production of α-galactosidase by SmF using soybean vinasse as substrate, a residue of the alcoholic fermentation of soybean molasses. *Lactobacillus agilis* was selected for presenting the highest enzymatic activity (11.07 $U.mL^{-1}$) after 144 h of fermentation using a medium with 30% soluble solids that represents a C:N relation of 9, and inoculum size of 25% (v/v).

Table 1. Lignocellulosic Enzymes Production using different fibrous materials

			Activities and specific activities			
Microorganism	**Substrate**	**Fermentation time**	**Cellulase** ($UFP.gds^{-1}$)	**Xylanase** ($U.gds^{-1}$)	**Laccase** ($U.gds^{-1}$)	**Reference**
Pleurotus ostreatus e *P. sojor-caju*	Wheat straw, Rice straw, Sesame straw, Sawdust, Viniculture residues	Until fructification	2.20 $U.g^{-1}$		5.5 $U.mg^{-1}$ ptn.	[126]
Trichoderma reesei	Wheat straw	20 days	1.8 (FPase)			[131]
Trichoderma harzianum	Palm oil extraction residues	15 days	8.2 (FPase) (lab scale) 10.1 (FPase).gds^{-1} (pilot scale)			[132]
T. reesei e *A. niger*	Rice straw and sugarcane bagasse	96 h	T. reesei: 299.55 (CMCase) A. niger: 135.44 (CMCase)			[62]
Neurospora crassa	Wheat meal and straw,	96 h	492.8 (CMCase)	297.8		[133]
Pleurotus ostreatus	Sugarcane bagasse	8 days	0.013 (FPase) 0.18 (CMCase)	5.79	0.04	[63]
Lentinus Pleurotus sp.	Wheat straw and wheat leaves	7 e 10 days	325 (CMCase) 55 (FPase)	546.3	14.25	[127]
Trametes trogii	Sawdust + malt extract	43 days	779.71 $UCMC.gds^{-1}$	1144.1	210.2	[134]
Pleurotus sp.	Sawdust	10 days			378 $U.L^{-1}$	[128]
P. ostreatus e *T. versicolor*	Wheat straw	21 days	P. ost. 50 $mUCMC.gds^{-1}$ T. vers. 428 $mUCMC.gds^{-1}$	98 $mU.gds^{-1}$	1360 $mU.gds^{-1}$ 73 $mU.gds^{-1}$	[105]
Piptoporus betulinus	Wheat straw	42 days	20.063 $mUCMC.gds^{-1}$	1.92 $mU.gds^{-1}$		[105]
Lentinus tigrinus	Wheat straw + Wheat meal	21 days + 90 days	21 days: 300 $UCMC.gds^{-1}$ 90 days: 1200 $UCMC.gds^{-1}$.	in 20 days: 200	in 20 days: 30	[129]
P. ostreatus	Wheat straw	100 days	14 days: 203 $mU.gds^{-1}$	14 days: 55 $U.mgds^{-1}$	98 days: 3538 $mU.gds^{-1}$	[135]
Pleurotus ostreatus, *Pleurotus sojor-caju*	Banana residues	40 days	0.1 $U.mg^{-1}$	0.18 $U.mg^{-1}$	1.6 $U.mg^{-1}$	[130]

OBS: UCMC: units of CMC as substrate; UFP: units of filter paper as substrate in enzymatic assays.

2.9. Proteases

Proteases catalyse peptide-bond hydrolysis and mediate protein function for a multitude of cellular processes [138]. Among all proteases, alkaline proteases are used as detergent additives and they accounted for 40% of the total worldwide enzymes sales and 89% of the total protease sale [139, 140, 141].

An alkaline protease produced by a thermophilic bacteria *Bacillus subtilis* DM-04 was studied by Mukherjee et al. [141] in SSF using potato peel (51.7% of carbon; 2.6% of nitrogen and 19.9% of C/N ratio) achieving 400 U.gds^{-1}. The higher alkaline protease (2382 U.gds^{-1}) was obtained with a mixture of potato peel and *I. cylindrical* in the proportion 1:1 in tray-type bioreactor recovered with aluminum foil, incubated at 50ºC during 24 h. Low levels of acid and alkaline proteases were produced during fermentation using residues (cassava peel, corn cob, corn husk, oat husk and sugar cane bagasse) by *Penicillium janthinellum* CRC 87M-115 [88].

Alkaline protease production under SSF was investigated using isolated alkalophilic *Bacillus sp.* and green gram husk as substrate. Maltose and yeast extract supplementation increased protease production. The fermentation was conducted at 33ºC, pH 9.0, and moisture content (140%) during 60 h [142]. Protease activity reached more than 35,000 U.g^{-1}.

A protease from marine fungus *Engyodontium album* was produced under SSF using wheat bran, moistened with seawater until 60% (w/v). 1 ml of conidial inoculum was mixed thoroughly, incubated at 28ºC during 120 h, achieving a production of 5106 U.gdm^{-1}. This is an active highly alkaline and termostable protease, which makes it applicable in detergent industry [143].

A mesophilic *M. bacilliformis* strain was applied in wheat bran for acid protease production under SSFat 24°C for 72–96 h. The leaching efficiency of sodium chloride solutions to recover the enzyme from the fermented mass (extraction) was higher than that of non-ionic detergents or distilled water [144].

A SSF was conducted with rice bran and *R. oligosporus* ACM 145F, incubated at 37°C for 72 h, achieved a maximal production of an acid protease (1.6 U.mL^{-1}) at pH 2.0, decreasing its activity in pH values above 5.0 [145].

Sandhya et al. [146] tested the agro-industrial residues (wheat bran, rice husk, rice bran, spent brewing grain, coconut oil cake, palm kernel cake, sesame oil cake, jackfruit seed powder and olive oil cake) that were screened for neutral protease production. The best substrate was wheat bran in SmF and SSF. In SSF, the best results (31.2 U.gdm^{-1}) were obtained in a medium having initial moisture content of 43.6%), incubated at 30ºC for 72 h.

Kumar et al. [147] studied the production of acid protease by *Synergistes sp.* using a source of myoproteins – animal fleshing. It is a major proteinaceous solid waste generated during the manufacture of leather (in tanneries). Its elemental composition is: C (36.14%), H (8.31%), N (11.6%) and S (0.66%). *Synergistes* sp. was selected from an anaerobic digester used for the treatment of tannery solid waste and was selected for enhanced production of an extracellular acid protease (350–420 U.mL^{-1}) using animal fleshing as substrate.

Pseudomonas aeruginosa was a strain isolated from the tannery wastewater for its ability to produce alkaline protease (1160–1175 U.mL^{-1}) [148].

Jatropha seed cake is generated by oil extraction from seeds of *Jatropha curcas.* The oil is used in biodiesel production and it is one of the major energy crops in Thailand, but the residue contains toxic constituents such as antineutrinos and phorbol esters, which prevents its use in animal feed. Then Thanapimmetha et al. [149] used the residue deoiled *Jatropha curcas* seed cake for protease production by *Aspergillus oryzae* in SSF at 30ºC, 84 h, 45% of initial moisture and 10% of inoculum size achieving after optimization studies from 3094 U.gds^{-1} to 14108 U.gds^{-1}.

Jatropha contains approximately 47% of crude fat, 25% of crude protein, 10% of crude fibre, 5% of moisture and 8% of carbohydrates. Deoiled Jatropha seed cake (from *Jatropha curcas* - pinhao manso) was assessed for its suitability as substrate for enzyme production by SSF. Solvent tolerant *Pseudomonas aeruginosa* PseA produced protease activity of 1818 U.g^{-1} of substrate at 50% substrate moisture, in 72 h and pH of 6.0. Enrichment with maltose and peptone also increased protease production achieving around (11474.5 U.g^{-1}) [150].

2.10. Lipases

Lipids are mainly available as triacilglicerydes and a smaller amount as long-chain fatty acids. The lipid fraction contains fats, oils and greases. Industrial activities that generate biggest amounts of lipid effluents are: edible oils, creameries, dairy products industries, slaughterhouses, wool scouring and domestic effluents and restaurants, mainly fast-food , kitchen activities, bakeries and beverages industries [151]. Lipids are compounds that form insoluble oil films in aquatic surfaces, which blocks oxygen diffusion and can promote aquatic death [152] if any treatment are conducted [151]. In effluent treatments stations, high lipid concentration stimulates biomass flotation, increasing time of treatment and flocculants consumption [153]. Table 2 shows some residues used for lipase production reported in literature.

Discharger options of lipids residues are: application in land by spray irrigation, but it has disadvantages as odour problem and gas emission; physical-chemical treatment with blood coagulants and flocculants, with the disadvantage of high costs; anaerobic treatments, that showed operational problems due to solidification at lower temperatures, clogging and biomass loss through reactor`s outflow. Due to difficulty of dealing with all those problems, lipase application is another option to slaughterhouses effluents treatment [154, 155, 156].

Those are the residues produced after seeds oil extraction. Edible oil cakes contain a high protein rate and are usually used for animal feeding. According to Ramachandran et al. [157], 17 million tonnes are produced in Brazil, Argentina and India. This waste is a potential substrate for lipase production due to high levels of residual lipid, protein and nitrogen and is interesting because of low costs [158, 159]. Mafakher et al. [159] has isolated from agro-industrial wastewater two potential strains of *Yaworria lipolytica.* The strain M1 was capable of producing lipases in different oils such as: crude oil, glycerol, olive oil, paraffin and kerosene. Olive oil showed best lipase activities, reaching 11 U.mL^{-1}.

Lanciotti et al. [160] also has used *Yaworria lipolytica* RO12 for lipase synthesis in oil mill wastewaters (OMW), achieving high lipase activity (1677 U).

Haba et al. [161] selected the most important lipase producer (*Pseudomonas* sp 3AT) using frying oil as substrate (2% of frying oil plus some salts and nutrients). The production of 2748 U.L^{-1} and also *Pseudomonas aeruginosa* ATCC 111 production (1703.8 U.L^{-1}) was verified in a mix (1/1) of sunflower and olive oils culture medium.

Mongkolthanaruk & Dharmsthiti [152] studied three bacterial strains: *Pseudomonas aeruginosa* LP602, *Bacillus* sp. B304 and *Acinetobacter calcoaceticus* LP009 in treatment of lipid rich wastewater containing 21,000 $mg.L^{-1}$ of lipids.

Yunchalard et al. [162] isolated 55 microorganisms from soil and 40 from wastewater samples using olive oil rhodamine agar plates. The strain WL8 was able to produce the relatively highest in its lipase activity at 1.21 $U.mL^{-1}$ in a liquid medium containing 2% olive oil as sole carbon source at 37ºC, pH 7.0 for 48 h.

Table 2. Lipase production in different residues in submerged fermentation

Substrate	Strain	Optimum Production Conditions	Lipase Activity	Reference
Tomato processing waste	unindentified	32 days	* 0.278 $U.mg^{-1}$	[167]
Slaughter house waste (Beef tallow) was melted and emulsified	*Pseudomonas gessardii*	T 37◦C pH 5.0	Ca^{2+} increases activity; K^+, Mg^{2+} , Na^+ and Fe^{2+} inhibits slightly *1120 $U.mg^{-1}$ ** 139 $U.mL^{-1}$	[168]
Olive mil waste water supplemented with NH_4SO_3 and yeast extract.	*Candida rugosa*	Olive mil wastewater (148 $g.L^{-1}$ solids), pH 5.6	3511 $U.L^{-1}$	[169]
Sewage sludge fats from slaughter houses mix (1:1) with wood chips	unindentified	40% fats	Higher lipase activity with 30% of fats ** 40 $U.g^{-1}$	[170]
Oil refinery residue	*Penicillium citrinum*	-	Best N source: NH_4Cl **6738 $U.L^{-1}$ *4.38 $U.mg^{-1}$	[171]

*Specific Activity; **Maximum Activity.

Kuforiji et al. [163] has studied a pool of enzymes using several agricultural wastes for cultivation of *Pleurotus tuberregium*. All wastes were treated with 2% $CaCO_3$. Lipases from the fruitbodies of *Pleurotus tuber-regium* showed higher activities compared to activities of sclerotia in cotton wastes and rice straws with activities 2.5 $mg.h^{-1}.mg^{-1}$ of protein and 3.0 $mg.h^{-1}.mg^{-1}$of protein respectively. For groundnut shells sclerotia showed best results with 5.8 $mg.h^{-1}.mg^{-1}$ of protein.

Cordova et al. [164] has used olive oil cake and sugar cane bagasse to produce lipases with *Rhizomucor pusillus* and *Rhizopus rhizopodiformis*. Residues were supplemented with bactopeptone, glucose and salts. Using bagasse alone lipase production were 1.73 $U.mL^{-1}$ for *Rhizomucor pusillus* and 0.97 $U.mL^{-1}$ for *Rhizopus rhizopodiformis.* When mixed (50/50) olive oil cake and bagasse *Rhizopus rhizopodiformis* produced 79.60 $U.g^{-1}$, equivalent to 43.04 $U.mL^{-1}$.

Castilho et al. [165] evaluated the SSF and SFm with *Penicillium restrictum* for lipase production. SSF occurred for 24 h in babassu cake oil or olive oil as substrates. SmF was performed for 64 h in a medium containing meat peptone, yeast extract and olive oil. SmF productivity (0.266 $U.mL^{-1}.h^{-1}$) was higher than SSF (0.242 $U.mL^{-1}.h^{-1}$). SSF process provided unitary product costs 47% lower than the lipase market price, due to the use of cheap raw material.

Table 3. Lipase production by SSF in different residues

Residue/ waste	Microorganism	Maximum Activity and Conditions	Reference
Fish processing waste	*Bacillus altitudinis* AP-MSU	2.964 $U.g^{-1}$ 5.260 $U.g^{-1}$ (50ºC, pH 78)	[172]
Grease waste	*Penicillium chrysogenum SNP5*	40.7 $U.mL^{-1}$	[154]
Grease waste	*Penicillium chrysogenum*	46 $U.mL^{-1}$ (7 days, 30ºC, pH 7)	[173]
Vegetable oil-refining waste	Microbial derived from wastewater sludges	120,000 $UA.g^{-1}$ (61-65ºC; pH 7.7-9)	[174]
Soybean meal	*Penicillium* P74	203.72 $U.g^{-1}$ (120 h ; C/N ratio 6.11)	[175]
Castor bean	*Penicillium simplicisimum*	44.8 $U.g^{-1}$	[176]
Babassu oil cake	*P. simplicisimum*	75.8 $U.g^{-1}$ (72 h, 35ºC, pH 7.0) at 50 ◦C,pH 4.0-5.0	[177]
Poultry slaughter house	*P. restrictum*	21.4 $U.g^{-1}$ (20 h , 35ºC, 70% humidity)	[156]
Wheat bran	*Aspergillus niger*	SSF (9.14 $U.g^{-1}$ dss) SmF (1.46 $U.mL^{-1}$)	[166]
Almond meal	*Rhizopus oligosporus*	48 $U.g^{-1}$ (30ºC, 48 h, pH 6.0)	[178]
Babassu oil cake	*Penicillium restrictum*	30.3 $U.g^{-1}$ (24 h; C/N ratio 14.1)	[179]
Gingelly oil cake, wheat bran and others	*A.niger* MTCC 2594	363.6 $U.g^{-1}$ (72 h)	[180]
Dry rice bran powder	*Candida rugosa*	37 $U.gdb^{-1}$ (1 $L.min^{-1}$), 30ºC; 48 h	[181, 182]

Faloni et al. [166] has studied *Aspergillus niger* in SSF containing wheat bran, glucose, olive oil and salts, and SmF containing glucose in concentrations ranging 0-2% and olive oil ranging 0-2%. SSF had a better performance; reaching 4.8 $U.mL^{-1}$ compared to 1.46 $U.mL^{-1}$ in SmF. Optimum conditions of SSF were: 65% moisture content, 1.5% glucose and 1.5% olive oil. Maximum activity was 9.14 $U.g^{-1}$.

CONCLUSION

This chapter showed several examples of the use of agroindustrial wastes in enzymes synthesis. Fermentation techniques, such as SSF and SmF have been employed for different processes reported in the literature in the last two decades. The composition, pre-treatments, nature of substrates, process's parameters, strain, physic-chemical conditions and availability of nutrients should be taken into consideration for selecting a particular production technique. Depending on the produced enzyme, there are critical factors affecting the process yield.

REFERENCES

[1] P. Nigam, D. Singh, *J. of Basic Microbiol.* 34, 405-414 (1994).
[2] A. Pandey, C.R. Soccol, D. Mitchell, *Proc. Biochem.,* 35, 1153–1169 (2000).
[3] C.R. Soccol, L.P.S. Vandenberghe, *Bioch. Eng. J.*, 13, 205-218 (2003).
[4] M.R. Spier, A.L.Woiciechowski, L.P.S. Vandenberghe, C.R. Soccol, *Int. J. Food Eng.*, 116, 1-21 (2006).
[5] A. Pandey, C. Webb, C.R. Soccol, C. Larroche, *Enzyme Technology*, Asiatech Publishers, New Delhi (2005).
[6] P. Manjunath, B.C. Shenoy; M.R. Rao, *J. Applied Biochem.*, 5, 235–260 (1983).
[7] B. Jin, H.J. Leeuwen, B. Patel, H.W. Doelle, Q. Yu, *Proc. Biochem.*, 34, 59–65 (1999).
[8] H . Anto, U.B. Trivedi, K.C. Patel, *Biores. Technol.,* 97, 2006, 97, 1161-1166.
[9] FAO (2011). Food and Agriculture Organization of the United Nations. Cutting food waste to feed the world. <http://www.fao.org/news/story/en/item/74192/icode/≥ *Access in 25 jul 2012.*
[10] Q. Wang, X. Wang, X. Wang, H. Ma, *Proc. Biochem.*, 43, 280–286 (2008).
[11] P. Selvakumar, L. Ashakumary, A. Pandey, *Biores. Technol.* 65, 83-85 (1998).
[12] P. Ellaiah, K. Adinarayana, Y. Bhavani, P. Padmaja, B. Srinivasulu, *Proc. Biochem., 38*, 615-620 (2002).
[13] R. Kammoun, B. Naili, S. Bejar, *Biores Technol.,* 99, 5602–5609 (2008).
[14] S. Ramachandran, A.K. Patel, K.M. Nampoothiri, F. Francis, V. Nagy, G. Szakacs, A. Pandey, *Biores. Technol.,* 93, 169–174 (2004).
[15] P. Shah, K. Bhavsar, S.K. Soni, J.M. Khire, *J of Ind. Microbiol. Biotechnol.*, 36, 373–380 (2009).
[16] D.B. Mitchell, K. Vogel, B.J. Weimann, L. Pasamontes, A.P.G.M Van Loon, *Microbiol.*, 143, 245–252 (1997).
[17] R. Kammoun, A. Farhat, H.Chouayekh, K. Bouchaala, S. Bejar, *Annals of Microbiol.,* 62, 155–164 (2012).
[18] Z.H. Wang, X.F. Dong, G.Q. Zhang, J.M. Tong, Q. Zhang, S.Z. Xu, *Waste Managem. Res.*, 29, 1262–1270 (2011).

[19] S. Tang, L. Boehme, H. Lam, Z. Zhang, *Biochem. Eng. J.*, 43, 157–162 (2009).
[20] K. Chantasartrasamee, D.I.N. Ayuthaya, S. Intarareugsorn, S. Dharmsthiti, S. *Proc. Biochem.*, 40, 2285–2289 (2005).
[21] A. Ebune, S. AI-Asheh, Z. Duvnjak, *Biores. Technol.,* 53, 7-12 (1995).
[22] K.M. Nampoothiri, G.J. Tomes, K. Roopesh, G. Szakacs, V. Nagy, C.R. Soccol, A. Pandey, *Appl. Biochem. Biotechnol.,* 118, 205-214 (2004).
[23] D.N.X. Salmon, L.C. Piva, R.L. Binati, C. Rodrigues, L.P.S. Vandenberghe, C.R. Soccol, M.R. Spier, *Bioproc. Biosyst.Eng.*, 35(7), 1067-1079 (2012).
[24] B.K. Bajaj, M.A. Wani, *Eng. in Life Sci.*, 11, 620–628 (2011).
[25] B. Bogar, G. Szakacs, J.C. Linden, A. Pandey, R.P. Tengerdy, *J. Ind. Microbiol. Biotechnol.,* 30, 183–189 (2003).
[26] Chadha, BS; Harmeet, G; Mandeep, M; Saini, HS; Singh, N. *World Journal of Microbiology & Biotechnology*, 20,105–109 (2004).
[27] H.K. Gulati, B.S. Chadha, H.S. Saini, *Folia Microbiol.*, 52, 491–497 (2007).
[28] T.N. Mandviwala, J.M. Khire, *J. Ind. Microb. Biotechnol.*, 24, 237–243 (2000).
[29] M. Papagianni, S.E. Nokes, K. Filer, *Food Technol. Biotechnol.,* 39 319–326 (2001).
[30] A. Farhat, H. Chouayekh, M. Farhat, K. Bouchaala, S. Bejar, *Molec. Biotechnol.*, 40, 127–135 (2008).
[31] A. Tahir, B. Mateen, S. Saeed, H. Uslu, *Micología Apl. Int.*, 22, 51-57 (2010).
[32] M.R. Spier, R.C. Fendrich, P.C. Almeida, M. Noseda, R. Greiner, U. Konietzny, A.L. Woiciechowski, V.T. Soccol, C.R. Soccol, *World J. Microbiol. Biotechnol.,* 27, 267–274 (2011).
[33] M.R. Spier, R. Greiner, J.A. Rodriguez-León, A.L. Woiciechowski, A. Pandey, V.T. Soccol, C.R. Soccol, *Food Technol. Biotechnol.,* 46, 178–182 (2008).
[34] M.L.C. Neves, M.F. Silva, C.M. Souza-Motta, M.R. Spier, C.R. Soccol, T.S. Porto, K. Moreira, A.L.F Porto, *Molecules,* 16, 4807-4817 (2011).
[35] B. Singh, T. Satyanarayana, *Appl.Biochem. Biotechnol.*, 133, 239-250 (2006).
[36] B. Singh, T. Satyanarayana, *Biores. Technol.*, 99, 2824–2830 (2008).
[37] D.E. Rodríguez-Fernández, J.A. Rodríguez-León, J.C. Carvalho, S.G. Karp, W. Sturm, J.L. Parada, C.R. Soccol, *Biores. Technol.*, 118, 603–606 (2012).
[38] A. Mittal, G. Singh, V. Goyal, A. Yadav, N.K. Aggarwal, *Innovative Romanian Food Biotechnol.*, 10, 18-27 (2012).
[39] A.I. El-Batal, H. Abdel Karem, *Food Res. Int.*, 34, 715–720 (2001).
[40] J.V. Madeira Jr, J.A. Macedo, G.A. Macedo, *Biores. Technol.*, 102, 7343–7348 (2011).
[41] P. Esakkiraj, G. Sandoval, S. Sankaralingam, G. Immanuel, A. Palavesam, *Annals Microbiol.*, 60, 461–468 (2010).
[42] Hong, K; Ma, Y; Li, M., 91-93, 777-785 (2001).
[43] Vassilev, N; Vassileva, M; Bravo, V; Serrano, MF; Nikolaeva, I. , *Industrial Crops and Products*, 26, 332–336 (2007).
[44] K. Roopesh, S. Ramachandran, K.M. Nampoothiri, G. Szakacs, A. Pandey, *Biores. Technol.*, 97, 506–511 (2006).
[45] S. Popanich, C. Klomsiri, S. Dharmsthiti, *Biores. Technol.*, 87, 295–298 (2003).
[46] R. Rani, S. Ghosh, *Biores. Technol.*, 102, 10641–10649 (2011).
[47] S. Bhargav, B.P. Panda, M. Ali, S. Javed, *Chem. Biochem. Eng. Quarterly*, 22, 49–70 (2008).

[48] S. Ramachandran, K. Roopesh, K.M. Nampoothiri, G. Szakacs, A. Pandey, *Proc. Biochem.*, 40, 1749–1754 (2005).
[49] UNICA – União da Agroindústria Canavieira de São Paulo. Available in: http://www.unicadata.com.br/listagem.php?idMn=72. Acessed in 23th July 2012.
[50] Renewable Energy Policy Network, 2006. <http://new.ren21.net/default.aspx>, Paris, France. Acessed in 12th Nov. 2012.
[51] R. Sanjuan, J. Anzaldo, J. Vargas, J. Turrado, R. Patt, *Holz als Roh-und Werkstoff*, 59, 447–450 (2001).
[52] S.E. Jacobsen, C.E. Wyman, *Ind. Eng. Chem. Res.*, 41, 1454–1461 (2002).
[53] R.C.L.B. Rodrigues, M.G.A. Felipe, J.B.A. Sil, M, Vitolo, M., *Proc. Biochem.*, 38, 1231–1237 (2003).
[54] C.R. Soccol, L.P.S. Vandenberghe, A.B.P. Medeiros, S.G. Karp, M. Buckeridge, L.P. Ramos, A.P. Pitarelo, V. Ferreira-Leitão, L.M.F. Gottschalk, M.A. Ferrara, E.P.S. Bon, L.M.P. Moraes, J.A. Araújo, F.A.G. Torres, *Biores. Technol.*, 101, 2011[a], 101, 4820-4825.
[55] J.R. Almeida, T. Modig, A. Petersson, B. Hahn-Hägerdal, G. Liden, M.F. Gorwa-Grauslund, *J. Chem. Technol. Biotechnol.*, 82, 340-349 (2007).
[56] A.S. Meyer, L. Rosgaard, H.R. Sørensen, *J. Cereal Sci.*, 50(3), 337-344 (2009).
[57] P. Giardina, G. Palmieri, B. Fontanella, V. Rivieccio, G. Sannia, *Arch. Biochem. Biophys.*, 376 (1), 171-179 (2000).
[58] F.D. Gago, Development of a bioprocess for celulases production under solid state fermentation using basidiomycete *Pleurotus* sp.: Production, characterization and purification studies. Master Science Thesis, Federal University of Paraná, Brazil, 2010 (*In Portuguese*).
[59] M. Karni, R.L. Deopukar, V.B. Rale, *World J. Microbiol. Biotechnol.*, 9, 475-478 (1993).
[60] E.M. Silva, A. Machuca, A.M.F. Milagres, *Proc. Biochem.*, 40, 161-164 (2005).
[61] Adsul, MG; Ghule, JE; Singh, R; Shaikh, H; Bastawde KB; Gokhale, DV; Varma, AJ. , *Carbohydrate Polymers*, 57, 67-72 (2004).
[62] R.K. Sukumaram, R.R. Singhania, G.M. Mathew, A. Pandey, *Renewable Energy*, 34, 421-424 (2009).
[63] I. Membrillo, C.Sánchez, M. Meneses, E. Favela, O. Loera, *Biores. Technol.*, 99, 7842-7847 (2008).
[64] T. Collins, C. Gerday, G. Feller, *FEMS Microbiol Rev.*, 29, 3–16 (2005).
[65] M.P. Coughlan, G.P. Hazlewood, *Biotechnol. Appl. Biochem.*, 17, 259–289 (1993).
[66] G.M. Maciel, L.P.S. Vandenberghe, R.C. Fendrich, B.E.D. Bianca, C.W.I. Haminiuk, C.R. Soccol, *Biotechno. Bioproc. Eng.*, 14, 748-755 (2009).
[67] R.S. Daniels,. Patent US 20120028801A1, 2012.
[68] T.J. Fang, C. Liao, S.C. Lee, *New Biotechnol.*, 27, 25-32 (2010).
[69] I. Seyis, N. Aksoz, *Food Tech. Biotechnol.*, 43, 37–40 (2005).
[70] S.H. Rose, W.H. van Zyl, *The Open Biotechnol. J.*, 2, 167-175 (2008).
[71] D.A. Bocchini, O.M.M.F. Oliveira, E. Gomes, R. da Silva. *Proc. Biochem.*, 40, 3653–3659 (2005).
[72] M. Meshram, A. Kulkarni, V.K. Jayaraman, B.D. Kulkarni, S.S. Lele, *Biochem. Eng. J.*, 40, 348-356 (2008).
[73] R. Kapilan, V. Arasaratnam, *Rice Science*, 18, 36–45 (2011).

[74] P.S. Murthy, M.M. Naidu. *World Appl. Sci. J.,* 8(2), 199-205 (2010).
[75] AACC (2000). Method 55-99: Barley Glossary. In Approved Methods, Z&h Edition. American Association of Cereal Chemists, St. Paul. MN.
[76] A. Haddar, D. Driss, F. Frikha, S. Ellouz-Chaabouni, M. Nasri, *Int. J. Biolog. Macromol.* 51(4), 647-56 (2012).
[77] Z.H. Xu, Y.L. Bai, X. Xu, J.S. Shi, W.Y. Tao, *World J. Microbiol. Biotechnol.,* 21, 575–581 (2005).
[78] M. Michelin, M.L.T.M. Polizeli, D.P. Silva, D.S. Ruzene, A.A. Vicente, J.A. Jorge, H.F. Terenzi, J.A. Teixeira, *J. Ind. Microbiol. Biotechnol., 38,* 1979–1984 (2011).
[79] M. Kapoor, L.N. Nair, K.R. Chnader, *Biochem. Eng. J.*, 38, 88-97 (2008).
[80] D.E. Rodríguez-Fernández, J.A. Rodríguez-León, J.C. de Carvalho, W. Sturm, C.R.Soccol, *Biores. Technol.* 102, 10657–10662 (2011).
[81] A.B. Díaz, I. Ory, I. Caro, A. Blandino, *Food Bioprod. Processing*, 90, 72–78 (2012).
[82] C. Paganin, N.C. Silva, A. Nogueira, G. Wosiacki. Ciência e Agrotecnologia, 29, 1231-1238 (2005).
[83] V.K. Joshi, D. Attri, *Natural Product Radiance*, 5, 289-296 (2006).
[84] G.S. Dhillon, S. Kaur, S.K. Brar, M. Verma, *Ind. Crops Prod.*, 38, 6–13 (2012).
[85] P. Yang, L. Guo, S. Cheng, N. Luo, J. Lin. *Renewable energy*, 36, 3268-3272, (2011).
[86] G. Mandalari, G. Bisignano, R.B. Lo Curto, K.W. Waldron, C.B. Faulds. *Bioresour. Technol.*, 99, 5130-5133 (2008).
[87] S.I. Mussato, I.C. Roberto, *J. Chem. Technol.Biotechnol.*, 81, 268-274 (2006).
[88] L.A. Oliveira, A.L.F. Porto, E.B.Tambourgi, *Biores. Technol.,* 97, 862–867 (2006).
[89] G.T. Dobrev, I.G. Pishtiyski, V.S. Stanchev, R. Mircheva, *Biores. Technol.*, 98, 2671–2678 (2007).
[90] Y. Su, X. Zhang, Z. Hou, X. Zhu, X. Guo, P. Ling *N. Biotechnol.*, 28, 40-46 (2011).
[91] J.X. Heck, L.H.B. Soares, M.A.Z. Ayub, *Enzyme Microb. Technol.*, 37, 417–423 (2005).
[92] A.A. Antoine, D. Jacqueline, P. Thonart, *Appl. Biochem. Biotechnol.*, 160, 50-62 (2010).
[93] S. Couri, S.C. Terzi, G.A. Saavedra Pinto, S.P. Freitas, A.C.A. Costa, *Proc. Biochem.*, 36, 255–261 (2000).
[94] F.M. Rombouts, W. Pilnik, *Proc. Biochem.*, 18, 9–13 (1978).
[95] M.R. Spier, R.C. Fendrich, P.C. Almeida, M. Noseda, R. Greiner, U. Konietzny, A.L.Woiciechowski, V.T. Soccol, C.R. Soccol, *World J. Microbiol. Biotechnol.*, 27, 267–274 (2011).
[96] E.S. Martins, D. Silva, R. Silva, E. Gomes, *Proc. Biochem.*, 37, 949–954 (2002).
[97] S.R. Patil, A. Dayanand, *Biores. Technol.,* 97, 2340–2344 (2006).
[98] Y. Lin, S. Tanaka, *Appl.Microbiol. Biotechnol.,* 69, 627–642 (2006).
[99] B. Ozturk, D. Cekmecelioglu, Z.B. Ogel, *J. Mol. Catalysis B: Enzymatic*, 64, 135–139 (2010).
[100] N. Mosier, C. Wyman, B. Dale, R. Elander, Y.Y. Lee, M. Holtzapple, M. Ladisch, *Biores. Technol.*, 96, 673–686 (2005).
[101] N. Beukes, B.I. Pletschke, *Biores. Technol.*, 102, 5207–5213 (2011).
[102] EMBRAPA. Soybean in numbers (Soja em numerous).< http://www.cnpso.embrapa.br /index.php?op_page=294&cod_pai=16> Acessed in 23th July 2012, (*In Portuguese*).

[103] POLINUTRI – Poli Nutri Alimentos LTDA. Soy bran: processing and quality, 2001. <http://www.polinutri.com.br/conteudo_ artigos_anteriores_janeiro.htm>. Acessed in 30th May 2009 (*In Portuguese*).
[104] L.P.S. Vandenberghe, C.R. Soccol, M.R. Spier, V.Weingartner Processo para produção de mananase de origem microbiana utilizando subprodutos/resíduos industriais (BR 1020120065649). 2012, deposited Mai 02th 2012
[105] V. Valášková, P. Baldrian, *Res. Microbiol.*, 157, 119–124 (2006).
[106] M. Wu, C. Tang, J. Li, H. Zhang, J. Guo, *Carbohyd. Res.*, 346, 2149–2155 (2011).
[107] H.M. Ferreira, E.X.F.Filho, *Carbohydrate Polymers*, 57, 23–29 (2004).
[108] Y. Basiron, *Eur. J. Lipid Sci. Technol.*¸109, 289–295 (2007).
[109] CMADS. Environmental Comission and Sustantaible Development.. Available in: <http://www.camara.gov.br/sileg/>, Accessed in Feb. 14, 2011 (*In Portuguese*).
[110] E.M. Düsterhöft, A.G.J. Voragen, F.M. Engels, *J. Sci. Food Agric.*, 55, 411–422 (1991).
[111] K.E.B. Knudsen, *Anim. Feed Sci. Technol.* 67, 319–338 (1997).
[112] F.O. Heidorne, P.O. Magalhães, A.L. Ferraz, A.M.F. Milagres, *Enz. Microb. Technol.* 38, 436–442 (2006).
[113] P.O. Magalhães, A.M.F. Milagres, *Int. Biodeter. Biodeg.*, 63, 191–195 (2009).
[114] F.G. Siqueira, E.G. Siqueira, P.M.D. Jaramillo, M.H.L. Silveira, A. Jürgen, F.A. Couto, L.R. Batista, E.X.F. Filho, *Int. Biodeter. Biodegrad.*, 64, 20–26 (2010).
[115] C. Xu, F. Ma, X. Zhang, *J. Biosci. Bioeng.*, 108(5), 372–375 (2009).
[116] T. Lin, C. Chen, C. *Proc. Biochem.* 39, 1103–1109 (2004).
[117] M. Yanhe, Z. Peijin, L. Hongcan, Patent CN 1266096, 2000.
[118] Y. Liu, Y. Li, Y. Wang, S. Xu, Y. Xu, Patent: CN 101481674, 2009.
[119] A. Kunamneni, S. Camarero, C. García-Burgos, F.J. Plou, A. Ballesteros, M. Alcade, *Microb. Cell Factories*, 7, 32 (2008).
[120] S.S. Kahraman, I.H. Gurdal, *Biores. Technol.*, 82, 215-217 (2002).
[121] S. Ntougias, P. Baldrian, C. Ehaliotis, F. Nerud, T. Antoniou, V. Merhautová, G.I. Zervakis, *Chemosphere*, 88, 620-626 (2012).
[122] H. Lakhtar, M. Ismaili-Alaoui, A. Philippoussis, I. Perraud-Gaime, S. Roussos, *Int. Biodeter. Biodegrad.*, 64, 167-172 (2010).
[123] S. Karp, Development of a biotreatment for delignification of sugarcane bagasse and production of laccases. *Doctorate thesis*, Federal University of Paraná, 2012.
[124] P. Dwivedi, V. Vivekanand, N. Pareek, A. Sharma, R.P. Singh, *New Biotechnol.*, 28, 616 (2011).
[125] S. Sen, P.P. Singh, V.S. Rathore, B.M.J. Pereira, *J. Scientific Ind. Res.*, 63, 739-746 (2004).
[126] S. Kurt, S. Buyukalaca, *Biores. Technol.*, 101, 3164-3169 (2010).
[127] V. Elisashvili, M. Penninckx, E. Kachlishvili, N. Tsiklauri, E. Metreveli, T. Kharziani, G. *Biores. Technol.*, 99, 457-462 (2008).
[128] M. Stajic, L. Persky, D. Friesem, Y. Hadar, S.P. Wassaer, E. Nevo, J. Vucojevic, *Enzyme Microb. Technol.*, 38, 65-73 (2006).
[129] B.E. Lechner, V.L. Papinutti, *Proc. Biochem.*, 41, 594-598 (2006).
[130] G.V. Reddy, P.R. Babu, P. Komaraiah, K.R.R.M. Roy, I.L. Kothari, IL. *Proc. Biochem.*, 38, 1457-1462 (2006).
[131] M. Lever, G. Ho, R.C. Ruwisch, *Biores. Technol.*, 101(18), 7094-7098 (2010).

[132] M.Z. Alam, A.A. Mamun, I.Y. Qudsieh, S.A. Muyibi, H.M. Salleh, N.M. Omar, *Biochem. Eng. J.*, 46, 61-64 (2009).
[133] I. Dogaris, G. Vakontions, E. Kalogeris, D. Mamma, D. Kekos, *Ind. Crops Prod.*, 29, 404-411 (2009).
[134] L. Levin, C. Herrmann, V.L. Papinutti, *Biochem. Eng. J.*, 39, 207-214 (2008).
[135] P. Baldrian, V. Valáskova, V. Merhautova, J. Gabriel, *Res. Microbiol.*, 156, 670-676 (2005).
[136] C.R. Soccol, A.L. Woiciechowski, M.R. Spier, A.B.P. Medeiros, A.K. Haghi.,In: AK Haghi (Org), Food Science: Applied Academic Press, (2011b).
[137] C.T.N. Sanada, S.G. Karp, M.R. Spier, A.C. Portella, P.M. Gouvêa, C.T. Yamaguishi, L.P.S. Vandenberghe, A. Pandey, C.R. Soccol. *Food Res. Int.*, 42, 476–483 (2009).
[138] I. Saska, D.J. Craik, *Trends in Biochem. Sci.*, 33(8), 363-368 (2008).
[139] R. Gupta, Q.K. Beg, P. Lorenz, *Appl. Microb. and Biotech.*, 59, 15–32 (2002).
[140] O. Krik, T.V. Borchert, C.C. Fuglsang, *Curr. Opin. Biotechnol.*, 13, 345–435 (2002).
[141] A.K. Mukherjee, H. Adhikari, S.K. Rai, *Biochem. Eng. J.*, 39, 353–361 (2008).
[142] R.S. Prakasham, C.S. Rao, P.N. Sarma, *Biores. Technol.*, 97, 1449–1454 (2006).
[143] S. Chellappan, C. Jasmin, S.M. Basheer, K.K. Elyas S.G. Bhat, M. Chandrasekaran, *Proc. Biochem.*, 41, 956–961 (2006).
[144] H.M. Fernández-Lahore, E.R. Fraile, O. Cascone, *J. Biotechnol.*, 62, 83–93 (1998).
[145] L. Ikasari, D.A. Mitchell, *Enz. Microb. Technol.*, 19, 171-175 (1996).
[146] C. Sandhya, A. Sumantha, G. Szakacs, A. Pandey, *Proc. Biochem.*, 40, 2689–2694 (2005).
[147] A.G. Kumar, N. Nagesh, T.G. Prabhakar, G. Sekaran, *Biores. Technol.*, 99, 2364–2372 (2008a).
[148] A.G. Kumar, S. Swarnalatha, B. Sairam, G. Sekaran, *Biores. Technol.*, 99, 1939-1944 (2008b).
[149] A. Thanapimmetha, B. Luadsongkram, P. Titapiwatanakun, Srinophakun, *Ind. Crops Prod.*, 37, 1-5 (2012).
[150] N. Mahanta, A. Gupta, S.K. Khare, *Biores. Technol.*, 99(6), 1729-1735 (2008).
[151] S. Kumar, A. Mathur, V. Singh, S. Nandy, S.K. Khare, S. Negi, *Biores. Technol.*, 120, 300-304 (2012).
[152] W. Mongkolthanaruk, S. Dharmisthiti, *Int.Biodeter. Biodegrad.*, 50, 101-105 (2002).
[153] A.A. Mendes, H.F. Castro, E.B. Pereira, A. Furigo Jr, *Química Nova*, 28 (2), 296-305 (2005).
[154] A.P. Kempka, *11th International Congress of Engineering and Food* (ICEF), 2011.
[155] D.I. Massé, L.Masse. *Canad. Agric. Eng.*, 42 (3), 139- 146 (2000).
[156] A.B.G. Valladão, D.M.G. Freire, M.C. Cammarota, *Int. Biodeter. Biodegrad.*, 60, 219–225 (2007).
[157] S. Ramachandran, K.S. Singh, C. Larroche, C.R. Soccol, A. Pandey. *Biores. Technol.*, 98, 2000–2009 (2007).
[158] A. Salihu, M.Z. Alama, M.I. Abdulkarima, H.M. Salleha, HM. *Resources Cons. Recyc.*, 58, 36–44 (2012).
[159] L. Mafakher, M. Mirbagheri, F. Darvishi, I. Nahvi, H. Zarkesh-Esfahani, G. Emtiazi, *New Biotechnology*, 27(4), 337-340 (2010).
[160] R. Lanciotti, A. Gianotti, D. Baldi, D; R. Angrisani, G. Suzzi, D. Mastrocola, M.E. Guerzoni, *Biores. Technol.*, 96, 317–322 (2005).

[161] E. Haba, O. Bresco, C. Ferrer, A.Marques, M. Busquets, A. Manresa, *Enzyme Microb. Technol.*, 26, 40–44 (2000).
[162] S. Yunchalard, T. Polkummak, W. Piyatherawong, *J. Biotechnol.*, 150S, S1–S576 (2010).
[163] O.O. Kuforiji, I.O. Fasidi, *Biores. Technol.*, 99, 4275–4278 (2008).
[164] J. Cordova, M. Nemmaoui, M. Ismaili-Alaoui, A. Morin, S. Roussos, M. Raimbault, B. Benjilali, *J. Molec. Catal. B Enzym.*, 5, 75–78 (1998).
[165] L.R. Castilho, C.M.S. Polato, E.A. Baruque, G.L. Sant'Anna Jr, D.M.G. Freire, *Biochem. Eng. J.*, 4, 239–247 (2000).
[166] G. Falony, J.C. Armas,J.C.D. Mendoza, J.L.M. Hernández, *Food Technol. Biotechnol.* 44 (2), 235–240 (2006).
[167] R. Sarada, R. Joseph. *Enzyme Microb. Technol.*, 15, 339-342 (1993).
[168] K. Ramani, L.J. Kennedy, M. Ramakrishnan, G.Sekarana, *Proc. Biochem.*, 45, 1683–1691 (2010).
[169] C. Gonçalves, M. Lopes, J.P. Ferreira, I.Belo, *Biores. Technol.* 100, 3759–3763 (2009).
[170] T. Gea, P.Ferrer, G. Alvaro, F. Valero, A. Artola, A. Sanchez, *Biochem. Eng. J.*, 33, 275–283 (2007).
[171] O.A. Miranda, A.A. Salgueiro, M.C.B. Pimentel, J.L. Lima Filho, E.H.M. Melo, N. Durfin, *Biores. Technol.*, 69, 145-147 (1999).
[172] P. Esakkiraj, R. Usha, A. Palavesam, G. Immanuel *Food Bioprod. Proc.* 90, 370–376 (2012).
[173] S. Kumar, N. Katiyar, P. Ingle, S. Negi, *Biores. Technol.* 102, 4909–4912 (2011).
[174] A. Santis-Navarro, T. Gea, R. Barrena, A. Sánchez, *Biores. Technol.* 102, 10080–10084 (2011).
[175] E. Rigo, J.L. Ninow, M. Di Luccio, J.V. Oliveira, A.E. Polloni, D. Remonatto, F. Arbter, R. Vardanega, D. Oliveira, H. Treichel, *LWT - Food Sci. Technol.* 43, 1132-1137 (2010).
[176] M.G. Godoy, M.L.E. Gutarra, O.L.T. Machado, D.M.G. Freire, *Enzyme Microb. Technol.* 44, 317–322 (2009).
[177] M.L.E. Gutarra, M.G. Godoy, F. Maugeri, M.I. Rodrigues, D.M.G. Freire, L.R. Castilho, *Biores. Technol.* 100, 5249–5254 (2009).
[178] I. Ul-Haq, S. Idrees, M.I. Rajoka, *Proc. Biochem.* 37, 637-641 (2002).
[179] A.K. Gombert, A.L. Pinto, L.R. Castilho, D.M.G. Freire, *Proc. Biochem.* 35, 85–90 (1999).
[180] N.R. Kamini, J.G.S. Mala, R. Puvanakrishnan *Proc. Biochem.*, 33, 505–511 (1998).
[181] P.V. Rao, K. Jayaraman, C.M. Lakshmanan, *Proc. Biochem.* 28, 385–389 (1993).
[182] P.V. Rao, K. Jayaraman, C.M. Lakshmanan, *Proc. Biochem.* 28, 391–395 (1993).

In: Enzymes in Value-Addition of Wastes
Editors: Satinder Kaur Brar and Mausam Verma
ISBN: 978-1-62948-968-1

Chapter 2

ENZYMES IN VALUE-ADDITION OF AGRICULTURAL AND AGRO-INDUSTRIAL RESIDUES

Neha Babbar and Harinder Singh Oberoi*
Central Institute of Post-Harvest Engineering and Technology,
P.O. PAU, Ludhiana, India

ABSTRACT

Enzymes play a pivotal role in most of the bioconversion and biotransformation processes. Agricultural or agro-industrial wastes are generally composed of residues, such as peels, pulp, stone, seeds, husk, stalks, straw, pods, shells, stubble, among others. Such wastes are generated in considerable amounts leading to their poor disposal and environmental pollution. The wastes generated during the processing of meat, poultry and fish are generally composed of bones, feathers, hide, hair etc. Though some of the residues, such as hide and bones mainly obtained from slaughter houses are employed for production of value-added products, still most of the residues are not put to any significant commercial use in many countries across the world. Some of the important value-added compounds that can be produced using such wastes are enzymes, *viz.*, cellulases, pectinases, proteases, phytases, xylanases, amylases; oligosachharides, such as galactooligosachharides (GOS), fructooligosaccharides (FOS), xylooligosachharides (XOS) and many other industrially important compounds, such as bioethanol, biobutanol, biohydrogen, xylitol, organic acids, vitamins, antibiotics and platform chemicals. Enzymes, such as amylases and isomerases are extensively used for production of high fructose corn syrup (HFCS) or concentrated sugar syrups. Pectinases in conjunction with cellulases find an important role in fruit juice extraction and clarification, while cellulases are indispensable in bioethanol production from lignocellulosic biomass, in addition to having an important role in textile industry, pulp and paper industry and formulation of animal feeds. Proteases have a substantial role in digestion of food, blood clotting, strengthening the body immune system and tenderization of meat. Most of these enzymes are a mixture of different enzyme components which have broad applications in food, pharmaceutical and textile industry. This chapter describes the processes for enzymatic

* Corresponding author's Tel: 91-161-2313126, Fax: 91-161-2308670, Email: hari_manu@yahoo.com.

bioconversion of wastes into important products having direct or indirect applications in food industry.

Keywords: Enzymes; Oligosaccharides; Agro-Industrial residues; Lignocellulosics; Biofuels; Fermentation; Phenolics

1. INTRODUCTION

"A thing is right when it tends to preserve the integrity, stability and beauty of the biotic community. It is wrong when it tends to do otherwise (Aldo-Leopold)"

Processing of fruits and vegetables generate a substantial amount of residues in the form of pods, peel, pulp, stones and seeds. The mass of by-products obtained as a result of processing tropical crops may approach or even exceed that of the corresponding valuable product affecting the economics of growing tropical exotic crops [1]. The fruit and vegetable solid waste contains, mainly soluble sugars and other hydrolysable materials and fiber. These residues are disposed off in the municipal bins or are left to rot because of the non-availability of proper infrastructure to handle such a huge quantity of biomass or/and an established commercial use for such wastes. Disposal of such wastes may present an added cost to processors and direct disposal to soil or landfill may cause serious environmental problems. Lignocellulosic waste is another agricultural waste which comprises of rice straw, wheat straw, sugarcane bagasse, sweet sorghum bagasse, cotton stalk etc., which is produced in bulk and agro-wastes, such as rice straw is burnt in the fields after crop harvest in many countries across South-East Asia, resulting in smoke clouds causing loss of biomass and environmental pollution [2].

Conversion of abundant agricultural waste biomass to biofuels, oligosaccharides, and other value-added products presents a viable option for reducing environmental pollution, improving energy security and reducing green house emissions. Hence, it is felt that processing waste utilization is a necessity and a challenge to the agro-food industry and needs first and foremost consideration to make units economically viable through the recovery of the value-added by-products. Therefore, the exploitation of agro-industrial wastes for the development of value-added products is a promising area of research. Value addition of wastes is near impossible without the intervention of enzymes which help in transformation of agro-industrial waste to some important value-added products. Enzymes hydrolyze cell wall components, thereby increasing cell wall permeability leading to production of the value-added components. Various enzymes, such as cellulases, pectinases and hemicellulases are often required to disrupt the structural integrity of the plant cell walls, thereby enhancing the extraction of the desired component. Majority of the biological conversion of agro-industrial processing wastes into value-added products have been carried through solid-state bioprocessing [3]. Solid-state bioprocessing consists of the utilization of water insoluble substrates for microbial growth and it is usually carried out in solid- or semi-solid systems in the absence or near absence of water [4]. Successful application of enzymes for the extraction of a variety of products, including pigments, sugars, phenols, oils, etc., indicates that enzymes are effective in efficient recovery of the above constituents [5]. Enzymes can be obtained from bacteria, fungi, animal organs, insects or other biological materials. It is important to understand the catalytic property, mode of action, optimal operational conditions and the type of enzyme or enzyme combination which could be most appropriate to use them effectively for biotransformation processes. This chapter focuses on the biotechnological potential of using and upgrading renewable resources-"leftovers"/ wastes , especially the agro-industrial processing residues (wastes) using enzymes for developing high value compounds using enzymes.

2. IMPORTANT ENZYMES FOR INDUSTRIAL APPLICATIONS

Enzymes are protein biomolecules which serve as catalysts to speed up the chemical reactions. Value-addition of agricultural wastes includes the biotransformation of wastes to a useful product and also the production of crude enzymes using biowaste as a substrate. The agricultural residues are generally composed of starch, cellulose, hemicellulose, lignin and pectin.

Cellulose is the major component of plant cell wall. It is a linear homopolymer of anhydroglucose units linked by β-1,4 glycosidic bonds. Other polysaccharides, such as hemicellulose, starch, pectin and lignin are associated with cellulose. Unlike cellulose, hemicellulose is a heterosaccharide composed of different sugar residues for example, D-galactan, D-mannan, D-xylan, L-arabinan etc. [6]. The principal pentose sugar in hemicellulose is xylose. The other common pentose sugar is arabinose, which is usually linked to xylose and galactose in arabinoxylan and arabinogalactan, respectively. The chemical structure of lignin is very complex. Lignin is an amorphous polymer consisting of phenylpropane units and their precursors are three aromatic alcohols (monolignols) namely p-coumaryl, coniferyl and sinapyl alcohols. The respective aromatic constituents of these alcohols in the polymer are called p-hydroxyphenyl, guaicol and syringyl moieties [7]. Lignin does not exist in plant tissue as an independent polymer but it is bonded with other polymers, *viz*., cellulose and hemicellulose forming complexes with them. The major obstacle for utilization of the biomass is the complete separation of lignin-carbohydrate complexes which shields cellulose from enzymatic hydrolysis. The extraction of lignin from herbaceous plants is of paramount importance and the utilization of extracted lignin can lead to the industrial production of valuable food and industrial products such as vanillin, ferulic acid, vinyl guaiacol, among others. The composition of cellulose, hemicellulose and lignin in some of the lignocellulosics is depicted in Table 1.

Table 1. Composition of Agro-industrial residues

Agro-industrial residues	Lignin (%)	Cellulose (%)	Hemicellulose (%)	References
Corn cobs	6.1	33.7	31.9	[96]
Sugarcane bagasse	25	50	26	[17]
Wheat straw	8.9	32.9	24	[97]
Rice straw	12.56	35.45	23.78	[98]
Sunflower seed hull	29.40	24.10	28.60	[99]

Table 2. Enzyme production using different agro-industrial wastes

Support	Microbial strain	Enzyme	References
Sugarcane Bagasse	*Trichoderma versicolor*, *Flammulina velutipes*; *Aspergillus niger*	MnP, laccase MnP; pectinolytic; cellulase, beta glucosidase	[59, 100, 101],
Wheat Bran	Ganoderma; *Aspergillus carbonarius*; *Aspergillus niger*, *Aspergillus niveus*	Laccases MnP; Pectinases; glucoamylase, catalase	[102, 103)
Wheat straw	*Phlebia radiate*, *Bacillus subtilis*	Lip, MnP, laccase; protease	[104]
Orange peelings, Citrus pulp bagasee	*Trametes hiruta*, *Thermoascus aurantiacus;* *Trichoderma vulgaris*, Polporous spp	Laccase, pectinases; Pectic enzyme cellulase & ligninase	[105, 106, 107]
Banana skin	*Trametes pubescences*; Cellulolytic and pectinolytic enzyme	Laccase	[108]
Grape pomace	*Aspergillus awamori*	Pectinases	[109]
Corn fiber xylan	*Fusarium proliferatum*	beta-xylosidase	[110]
Cauliflower waste	*Aspergillus niger*	Glucoamylase	[111]
Soybean hulls supplemented with wheat bran	*Trichoderma* and *Aspergillus*	Cellulolyic enzyme	[112]
Apple pomace	*Aspergillus niger*	Cellulases and pectinases beta-fructofuranosidase Ethanol	[35]

Pectins are polymers of D-galactopyranosyluronic acids joined by α-D-(1→4) glycosidic linkages. Pectic substances are ubiquitous in plants and are the major components of the middle lamella, thus constituting the majority of fruit processing wastes (especially from citrus fruits) and representing a potential field of application for pectinases. Based on the type of modifications of the backbone chain, pectic substances are classified into protopectin, pectic acid, pectinic acid and pectin [8]. Protopectin is a parent pectic substance and upon hydrolysis yields pectin and pectinic acid. Pectic acids are the galacturonans that contain negligible amounts of methoxyl groups. Normal or acid salts of pectic acid are called as pectates. Pectinic acids are the galacturonans with varying amounts of methoxyl groups. Pectinic acid alone has the unique property of forming a gel with sugar and acid. Pectin is a generic name for the mixture of widely differing compositions containing pectinic acid as the major component.

Starch contains the glucose linked up by the α-glucosidic bonds. Starch consists of two distinct fractions (a) amylase linear α-1,4-linked glucans and (b) amylopectin-linear α-1,4-linked glucans branch with α-1,6 linkages. The following enzymes are needed to degrade these components so that the wastes could be transformed into value-added products.:

2.1. Cellulolytic Enzymes

Cellulases play a significant role in emerging bio-energy and various other biotechnological processes, such as food, textile, paper and pulp, agriculture and environment. Cellulolytic enzymes are the most important industrial enzymes due to their versatile applications in various industries. The resurgence in utilization of biomass for bioethanol production and other value-added organic compounds has attracted major attention of researchers globally towards cellulases. Cellulase is not a single enzyme but is a complex of three major types of cellulases: endoglucanases, exoglucanases and cellobiase. Other names for endoglucanases are endo-1,4-β-glucanase, carboxymethylcellulases (CMCase), endo-1,4-β-D-glucanase. Endocellulases (EC 3.2.1.4) randomly cleaves internal bonds at amorphous sites that creates new chain ends. Exocellulases (EC 3.2.1.91) cleaves two to four units from the ends of the exposed chains produced by endocellulase, resulting in tetrasaccharides or disaacharides such as cellobiose. Cellobiase or β-glucosidase (EC 3.2.1.21) hydrolyses the exocellulase product into individual monosaccharides (Figure 1.). Bioconversion of lignocellulosic waste to soluble sugars relies on the synergistic action of complete cellulase system comprising of CBH, EG and BG [9].

2.2. Hemicellulases

A spectrum of hemicellulase enzymes synergistically acts on hemicellulose leading to its complete hydrolysis. These enzymes include endo-xylanase (endo-1,4-β-xylanase, E.C.3.2.1.8), β-xylosidase (xylan-1,4- β-xylosidase, E.C.3.2.1.37), α-glucuronidase (α-glucosiduronase, E.C.3.2.1. 139), α-arabinofuranosidase (α-L-arabinofuranosidase, E.C.3.2.1.55) and acetylxylan esterase (E.C.3..1.1.72). Among them, endoxylanases and β-xylosidases (collectively called as xylanases) are the two key enzymes responsible for hydrolysis of xylan, the major component of hemicellulose. Endo-xylanases act on homopolymeric backbone of 1,4-linked β-D-xylopyranose producing xylo-oligomers [10] while β-xylosidases act on these xylo-oligomers releasing xylose [11]. In nature, hemicellulases are produced by a wide variety of microorganisms, including bacteria and fungi [10].

2.3. Proteases

Proteases are enzymes that are able to hydrolyze peptide bonds. They can act near the ends of polypeptide chains (exopeptidases) or within them (endopeptidases) [12]. Exoproteases have been differentiated according to their substrate specificity as aminopeptidases, which are able to cleave peptides at the N-terminus and carboxypeptidases, which degrade peptides at the C-terminus. Endoproteases are classified according to their catalytic mechanism, which implies specificity in the enzyme active site. Proteases constitute the most important group of industrial enzymes currently in use.

2.4. Lipases

Lipases (triacylglycerol acylhydrolases) are water soluble enzymes and catalyze esterification, interesterification, acidolysis, alcohaolysis and aminolysis in addition to the hydrolytic activity on triglycerides [13]. Lipases have also been explored for the novel biotechnological applications such as synthesis of biopolymers and biodiesel, the production of enantiopure pharmaceuticals and flavour compounds [14]. Lipases are largely produced from filamentous fungi, such as *Penicillium chrysogenum*, *Penicillium cyclopium* etc. and bacterial strains like *Bacillus subtilis, Pseudomonas* [15].

2.5. Pectinases

Pectinases are responsible for the degradation of the long and complex molecules called pectins that occur as structural polysaccharides in the middle lamella and the primary cell walls of young plant cells. Pectinolytic enzymes can be applied in various industrial sectors wherever degradation of pectin is favorable for a particular process. The three major types of pectinases are: (i) pectinesterases also known as pectinmethyl hydrolase, which catalyzes de-esterification of the methoxyl group of pectin forming pectic acid and; (ii) depolymerizing enzymes which includes, (a) polymethylgalacturonases (PMG). They may be endo- PMG which causes random cleavage of α-1,4-glycosidic linkages of pectin, preferentially highly esterified pectin and exo- PMG causes sequential cleavage of α-1,4-glycosidic linkage of pectin from non-reducing end of the pectin chain ;(b) polygalacturonases (PG) catalyze the hydrolysis of α-1,4-glycosidic linkages in pectic acid (polygalacturonic acid). They are also of two types: Endo PG catalyses random hydrolysis of α-1,4-glycosidic linkages in pectic acid and Exo PG catalyzes hydrolysis in a sequential fashion of α-1,4-glycosidic on pectic acid (c) cleaving α-1,4-glycosidic α-1,4-glycosidic linkages by trans-elimination includes polymethylgalacturonate lyases (PMGL) that catalyses breakdown of pectin by trans-eliminative cleavage. They are endo-PMGL which catalyses random cleavage of α-1,4-glycosidic linkage in pectin and exo-PMGL that catalyses stepwise breakdown of pectin by trans-eliminative cleavage (d) polygalacturonate lyases which catalyses cleavage of α-1,4-glycosidic linkage in pectic acid by trans-elimination and includes endo-PGL which catalyses random cleavage of α-1,4-glycosidic linkages in pectic acid and exo-PGL that catalyzes sequential cleavage of α-1,4-glycosidic linkages in pectic acid (iii) protopectinase: This enzyme solubilizes protopectin forming highly polymerized soluble pectin (Figure 2).

2.6. Amylases

Enzymes responsible for the hydrolysis of starch are called amylolytic enzymes or simply amylases. Amylolytic enzymes form a large group of enzymes, among which the most common and the best known are α-amylases, β-amylases and glucoamylases. Since, starch is an essential source of energy, amylolytic enzymes are produced by a great variety of living organisms. Although, different amylases mediate the same reaction-they all catalyze the cleavage of the α-glucosidic bonds in the same substrate, while structurally and mechanically, they are quite different. Since 1950s, fungal amylases have been used to manufacture sugar syrups containing specific mixtures of sugars that could not be produced by conventional acid hydrolysis of starch. Alpha-amylases catalyses cleavage of α-1,4-glycosidic bonds in the inner region of the molecule, hence causing a rapid decrease in substrate molecular weight and viscosity. These endo-acting enzymes can be divided into liquefying or saccharifying α-amylases which preferentially degrade substrates containing more than fifteen or four glucose units, respectively. Two types of exo-acting amylases are commonly used for starch saccharification: β-amylases and glucoamylases. Both act on glycosidic linkages at the non-reducing ends of amylose, amylopectin and glycogen molecules, providing low molecular weight carbohydrates in the β-anomeric form. The main end product of hydrolysis catalysed by β-amylases is maltose, while glucoamylase generates glucose.

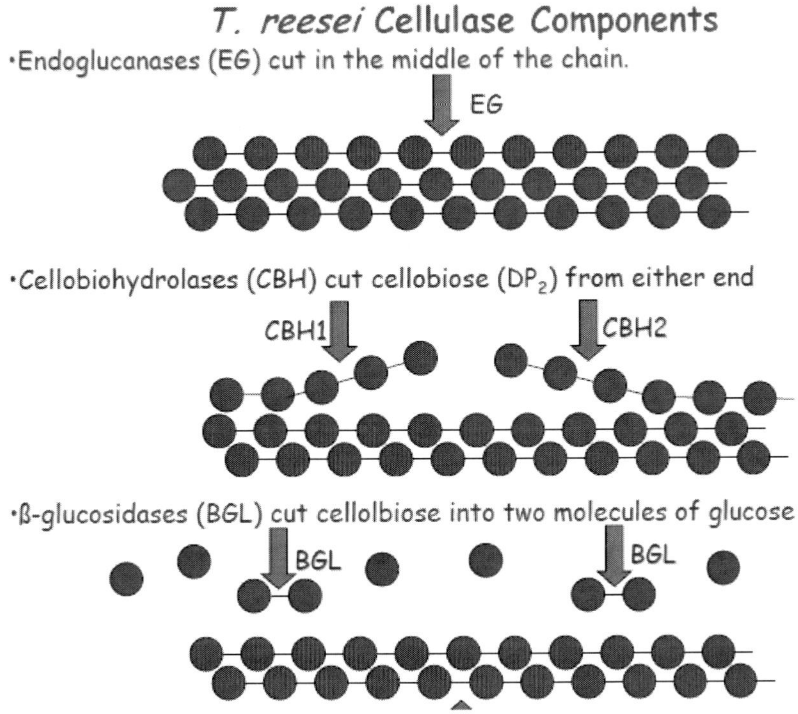

Figure 1. Bioconversion of cellulose to glucose using cellulases.

2.7. Ligninolytic Enzymes

The term lignin degrading enzymes encompasses mainly three oxidative enzymes; lignin peroxidase (LiP), manganese peroxidase (MnP) and laccase. The demand for these enzymes has increased in the recent years due to their potential applications in diverse biotechnological areas. Lignin degrading enzymes are widely used in the pollution abatement, especially in the treatment of industrial effluents that contains hazardous compounds, such as dyes, phenols and other xenobiotics. The industrial preparation of paper requires separation and degradation of lignin in wood pulp. Pretreatment of wood pulp with lignolytic enzymes might provide milder and cleaner strategies of delignification. Ligninolytic enzymes are commonly used for bio bleaching of kraft pulp. The ever increasing demand for these enzymes in the industrial sector necessitates the production of enzymes from inexpensive raw materials as well as utilization of these enzymes for production of value added products is discussed in the following section.

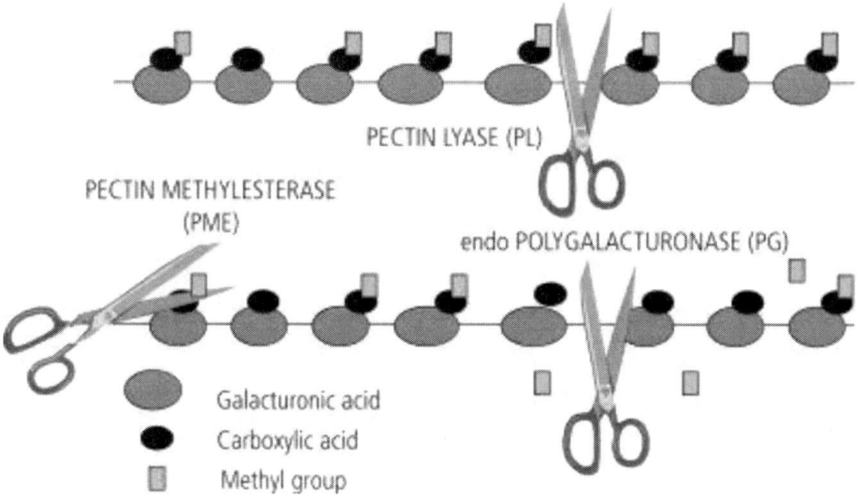

Figure 2. Mode of action of pectinases.

3. Enzyme Production through Fermentation

Enzymes are among the most important products obtained for human needs through microbial sources. Enzyme production is achieved using both solid-state fermentation (SSF) and submerged fermentation (SmF).

3.1. Solid-state Fermentation

Solid-state fermentation is defined as any fermentation process performed on a non-soluble material that acts both as physical support and source of nutrients in absence of free flowing liquid [16]. The low moisture content means that fermentation can only be carried out by a limited number of microorganisms, mainly yeasts and fungi, although some bacteria have also been used [17]. The nature of the solid substrate employed is the most important factor affecting SSF processes and its selection depends upon several factors mainly related with cost and availability and, thus, may involve the screening of several agro-industrial residues. In a SSF process, the solid substrate not only supplies the nutrients to the culture but also serves as an anchorage for the microbial cells. Among the several factors, which are important for microbial growth and activity in a particular substrate, particle size and moisture level/water activity are the most critical [18]. Agro-industrial residues have been widely used due to their potential advantages for filamentous fungi, which are capable of penetrating into the hardest of these solid substrates, aided by the presence of turgor pressure at the tip of the mycelium [19]. In addition, the utilization of these agro-industrial wastes on one hand provides alternative substrates and, on the other, helps in solving pollution problems caused by their disposal [13]. Some of the enzymes produced using different microbial strains *via* SSF are presented in Table 2.

3.2. Submerged Fermentation

Submerged fermentation involves inoculation of the microbial culture into the liquid medium for production of the desired product. Most of the commercial products are produced through the submerged fermentation processes. Fermentation processes are broadly classified into aerobic and anaerobic processes. Antibiotics and enzymes are produced through aerobic fermentation which involves the incorporation of oxygen into the liquid medium, while butanol production proceeds through the anaerobic mode, wherein addition of oxygen is detrimental. Certain fermentation processes like ethanol production involves facultative anaerobic organisms like *Saccharomyces cerevisiae* which can grow in the presence of oxygen leading to cell biomass production, but switch over to the anaerobic route during the fermentation stage. However, enzymes like amylases and proteases are generally produced through submerged fermentation process employing the aerobic fermentation route. Submerged fermentation processes can be easily scaled up with ease of automation and do not suffer from the heat mass transfer limitations. However, the major drawbacks associated with the submerged fermentation processes are the low productivity, high production cost and complexity of the medium.

4. Potential of Fruits and Vegetable Solid Waste for Bioconversion

In the horticultural sector, there has been a growth in both acreage and production to fulfill the requirements of global food demand. This intensity of production generates large amounts of plant by-products. The wastes/ by-products obtained from fruits and vegetables present a major disposal problem for the concerned industry. Fruits from the temperate zone are usually characterized by a large edible portion and moderate amounts of waste materials, such as peels, seeds and stones. In contrast, considerably higher ratios of by-products arise from tropical and sub-tropical fruit processing. It has been reported that the by-products of tropical exotic fruits contain high levels of various health enhancing substances that can be extracted from by-products to provide nutraceuticals [20]. Biological conversion processes of fruit processing wastes into various value-added products through solid-state fermentation have been of major interest to many laboratories around the world. Many of the potential products from fruit and vegetable residues have been developed, and such products include ethanol, methane, acetic acid, citric acid, mushrooms, enzymes and food ingredients [4]. Processing of fruits and vegetables results in different kinds of wastes, such as, citrus peel and pomace, apple pomace, banana peel, grape pomace and skin, apricot kernel, mango seed and pulp, litchi peel and pericarp, potato peel, tomato skin and pomace, carrot peel, pea pods, cauliflower florets and curds etc.

4.1. Citrus Waste

Citrus is one of the most important fruit crops of the world. The annual worldwide production of citrus fruits is about 122 million tons [21] and one third of the fruits are processed. Citrus fruits are processed, mainly to obtain juice, jams and squashes [22]. Worldwide industrial citrus wastes are estimated to be more than 15 million tons [23]. Citrus waste consisting of peel, pulp and seeds corresponds to half of the fresh fruit weight. Citrus wastes contain different carbohydrate polymers, which makes them interesting sources for the production of variety of products. Orange waste contains 11.8% fibre, 6.4% protein, 63% nitrogen, 6.7 % ash, 19% total sugars and 0.1 % pectin [24]. Kinow fruit, a part of the citrus family of fruits, known for its high juice content and characteristic juice flavor also generates a substantial quantity of wastes in the form of peels, pulp and seeds. Processes for production of cellulases and bioethanol through simultaneous saccharification and fermentation have been reported previously [25, 26]. Production of pectinases and cellulases using citrus waste is an alternative practical use of waste for the development of value- added products, such as enzymes [25, 27]. Discarded citrus pulp has been successfully used as an animal feed after its treatment with commercial pectinolytic enzymes. The end product showed good in-vitro digestibility and its protein content was comparable to agro-industrial waste products currently used as animal feed [28]. Phenolic compounds are also present in abundant quantities in citrus wastes, but the extraction efficiency of phenolic compounds may be associated with the factors such as number of hydroxyl groups and the site of binding as well as mutual position of hydroxyls in the aromatic ring [29]. Enzymatic hydrolysis of these phenolic glycosides seems to be a promising way to increase the concentration of free phenolics, and enhance the nutraceutical activity of such compounds [5]. Cellulolytic enzymes are extensively used for the extraction of phenolic compounds from different citrus peels [30].

4.2. Apple Residues

Apple is probably the oldest fruit known to mankind. Apple and apple products are one of the major fruit and fruit products consumed all over the world. The global annual production of apple is about 6.95 million tons [21]. Seventy per cent of the apple fruit is consumed for table purpose, while 20% is processed into value-added products, of which 65% is processed into apple juice, apple wine, cider, purees, jams and dried apple products. Apple pomace is the solid residue that remains after the extraction of juice from apple. Apple pomace contains large amount of water and therefore causes serious disposable problems. India produces about 25% of the total apple produced in the world [21]. In large scale processing industries, the wastes from apple can be categorized into two types. The first type is the fruit discarded into the sorting belt due to its partially bruised/spoiled nature and named as belt rejection. The second type is the apple pomace obtained after juice extraction. Owing to the high carbohydrate content, apple pomace is used as a substrate for the production of organic acids, enzymes, single cell protein, ethanol, low alcoholic drinks and pigments [31]. Apple pomace typically contains 66.4-78.2% (wb) moisture and 9.5-22% carbohydrates [32]. It also contains 26.4% dry matter, 4% proteins, 3.6% sugars, 6.8% cellulose, 0.38% ash, 0.42% acid and 8.7 mg calcium, per 100 g of wet apple pomace [33]. Numerous enzymatic transformations of apple pomace have been proposed for obtaining valuable products. Apple pomace has also been exploited for ethanol production after enzymatic hydrolysis using cellulases and pectinases [34]. Villas-Boas et al. [35] utilized *Candida utilis* to produce lignolytic enzymes and cellulolytic enzymes using apple pomace as a substrate under solid-state conditions. Apple pomace has been exploited for the production of polygalacturonase by *Lentinus edodoes* through solid-state fermentation [5].

4.3. Other Fruit Residues

The waste generated from the processing of papaya, pineapple and mango represent approximately 10-60% of the total fruit weight. Grapefruit peel waste was exploited for ethanol production after hydrolyzing cellulose, pectin, hemicellulose by pectinase and cellulase enzymes into monomeric sugars [36]. Pineapple waste is a by-product of the pineapple processing industry and consists basically of the residual pulp, peels and skin. It is not considered attractive as an animal feed, since it has a high fibre and low protein content. Ketnawa et al. [37] extracted bromelin from pineapple processing waste. Pineapple waste contains sufficient quantities of simple and complex sugars that have been used for bioethanol production. Krishna and Chandrasekaran [38] used banana fruit stalk as a substrate for α-amylase production by *Bacillus subtilis*. Oberoi et al. [39] reported 28.2 g/l ethanol from banana peel after its enzymatic hydrolysis. Commercial cellulase, β-glucosidase and pectinase at 8 FPU/g cellulose, 15 IU/g cellulose and 66 IU/g pectin, respectively hydrolyzed banana peels resulting in glucose and reducing sugars concentrations of 28.2 and 48 g/l respectively [40]. Grapes belong to the world's largest fruit crops with a global production of about 67 million tons [21]. About 80% of the total grapes produced are used in winemaking. The by-products obtained from grapes are seeds and pomace left after extraction of juice. A process for enzyme assisted extraction of polyphenols from grape pomace has recently been developed on pilot plant scale. Introduction of a 120 min enzymatic step during treatment of pomace resulted in 65.8% increase in anthocyanins yield [41].

4.4. Tomato Waste

Tomato contains a wide variety of antioxidants including vitamin E, ascorbic acid, carotenoids, flavonoids and phenolics [42]. Among them, lycopene, the carotenoid responsible for the deep red colour of ripe tomatoes has attracted considerable attention in recent years for its possible role in disease prevention. Tomato pomace i.e. the solid residue remaining after the industrial processing of tomatoes, is considered to be the one of the most promising candidate materials for production of lycopene. It consists of tomato peels and seeds, in a proportion depending on the specific process from which it is generated. Moreover, tomato pomace is a very rich source of lycopene, since this carotenoid progressively accumulates in the outer skin layer of tomatoes during the ripening process. At the end of ripening, lycopene levels in the skin are about five times higher than in the pulp [43]. Despite these advantages, lycopene recovery from this material is not straightforward, as revealed by the low yields achievable with conventional solvent extraction procedures [44]. The reasons are the compactness of the plant tissue, which hinders solvent penetration and transport to the lycopene containing chromoplasts, and the possible degradation of the carotenoid during recovery. However, an important point emerging from these studies is that the enzyme assisted extraction of lycopene is successful.

5. By- products of Fat and Oil Industry

Cooking oil waste is one of the major concerns of environmentalists because it is discharged into drains and other water bodies without treatment which, hitherto can be used for the production of biodiesel. Biodiesel has drawn attention as a non-toxic, biodegradable and renewable source of energy and other benefits includes lower exhaust emissions of particulate matter and greenhouse gases, such as CO and CO_2. For the production of biodiesel, lipases are considered as versatile catalysts that hydrolyze the ester bonds in tri-, di- and mono-aylglycerols. *Rhizopus oryzae* lipase was used for the production of fatty acid methyl esters using vegetable oil extracted from waste bleaching earths which contain approximately 40% of its weight as oil [45]. Transformation of waste oil to biodiesel by immobilized lipase from *Penicillium expansum* and *Thermomyces lanuginosus* was achieved by Li et al. [46] and Dizge et al. [47]. They also suggested that waste oil with high acid value is more promising feedstock for biodiesel production than refined oil, because of its low cost and high methyl ester yield. Another important use of oil processing industry waste is incorporation of deoiled cakes in animal feed. The endo-xylanase from *Neocallimastix patriciarum* was added to the deoiled canola seeds to enhance its digestibility in the rumen of cattle thereby, reducing the feed conversion ratio [48]. Lower feed conversion ratio means a better feed digestibility [49].

6. By-products of Sugar Processing Industry

The world sugar market is currently suffering from very high stocks and low prices. Therefore, long term options for the utilization of by-products are urgently required. Sugarcane (*Saccharum officinrum*) and sugarbeet (*Beta vulgaris*) are the most important crops for the production of sugar. Sugarcane bagasse, molasses and sugarbeet pulp are the important by-products of sugar processing plant. Owing to high cellulose content in sugarcane bagasse, it has been successfully used for the production of ethanol after enzymatic breakdown of cellulose to glucose which is further fermented by yeast to ethanol. Sugarcane bagasse has also been used for production of enzymes [50]. Sugarbeet pulp has an exclusive use in the production of ferulic acid, oligosaccharides etc. [51, 52].

6.1. Sugarbeet Pulp

Dried sugarbeet pulp is composed of 22-24 % cellulose, 21.5-23 % uronic acids, 1-2 % lignin, 7-8 % protein, 7.5-12 % ash and 0.5 % residual sucrose [53]. The primary technical challenge for the production of value-added products from sugarbeet pulp depends on bioconversion of above constituents by enzymes. Cocktail of various cellulolytic enzymes has been successfully used to recover ferulic acid, vanillin, diferullic acid, galacturonic acid from sugarbeet pulp [54, 55, 56]. Vanillin is usually synthesized from petrochemical substances, such as guaiacol or lignin. The microbiological production is based on the precursors eugenole or ferulic acid. Ferulic acid is found associated with cell walls of very few dicots, including sugar beets (0.36% related to dry weight). Sugarbeet pulp has also been exploited for the production of pectin oligosaccharides, feruloylated oligosaccharides using mixtures of cellulases and pectinases [52]. Ethanol production from sugarbeet pulp was evaluated after hydrolysis of sugarbeet pulp with fungal enzymes followed by fermentation using *Kluveromyces oxytoca* which resulted in 18.3g/l [57]. Sugarbeet pulp was also used as a substrate for pectinase production by *A. niger* [58].

6.2. Sugarcane Bagasse

Bagasse which results from the milling of sugar cane is among the world's most widely used and available non wood fibres. It contains 50% cellulose, 25% hemicellulose and 25% lignin [17]. The utilization of this lignocellulosic residue for the production of various industrially important enzymes has been a field of interest to the researchers. The fibrous nature of bagasse makes it more suitable for solid-state fermentation technique. The production of laccase and manganese peroxidase in SSF by *Trametes versicolor* and laccase alone by *Flammulina velutipes* using bagasse has been reported [59]. Manonmani and Sreekantiah [60] reported citric acid production using an enzymaticaly treated hydrolysate from alkali-treated bagasse followed by SSF. Cellulases, xylanases and β-glucosidase and ferulic acid esterase were produced by *Trichoderma reesei* and *Aspergillus awamori* which acted synergistically to poduce the above enzymes [50].

6.3. Molasses

Molasses is produced from both cane and beet, the product is the mother liquor separated from the crystallized sucrose. Cane molasses served as an excellent carbon and energy source for the economical production of glucoamylase by alginate immobilized *Thermomucor indicae* [61].

7. Agricultural Residues

7.1. Potential of Rice Straw as a Feedstock

Rice is the world's second largest cereal crop after wheat and produces the largest amount of crop residues. There are several reports on the use of rice straw for the production of enzymes, ethanol and organic acids [2]. Solid-state fermentation can be proposed as a better technology for commercial production of cellulases considering the low cost input and ability to utilize naturally available sources of cellulse as substrates. Utilization of abundant rice straw can help to reduce cellulase prices [62]. Cellulases are produced by many microorganisms, such as fungi, bacteria and actinomycetes. However due to high yields, fungi have been commercially exploited for the production of these enzymes. Among fungi, *Trichoderma* and *Aspergillus* have been exclusively studied as a microbial source of extracellular cellulase capable of hydrolyzing native cellulose [63]. Several species of *Aspergillus* show a high degree of β-glucosidase activity and synergism with *Trichoderma* cellulase enzymes [64]. The production of all the three lignocellulolytic enzymes by *Streptomyces psammoticus* using rice straw in submerged fermentation (SmF) has been reported [65]. Rice straw and wheat bran in the ratio 3:2 has been exploited for the production of cellulases using *Aspergillus niger* and *Trichoderma reseei*. Inclusion of wheat bran in the medium helps in enhancing cellulase production [66].

Cellulosic biofuels have many advantages over both gasoline and corn ethanol. Burning cellulosic ethanol rather than gasoline, for instance, could cut cars' greenhouse-gas emissions by 87 percent; corn ethanol achieves reductions of just 18 to 28 percent. Production of ethanol from rice straw has been reported by many researchers [2, 41, 67]

7.2. Potential of other Agricultural Residues and By-products

Sangeetha et al. [68] studied the production of fructosyl transferase by *A. oryzae* employing a great variety of agricultural by-products as substrates: Cereal brans (wheat bran, rice bran and oat bran), corn products (corn cob, corn bran, corn germ, corn meal, corn grits and whole corn powder), coffee and tea-processing by-products (coffee husk, coffee pulp, spent coffee and spent tea), sugarcane bagasse and cassava bagasse. In a study, a process for producing xylooligosaccharides (XOS) from corncob was developed. The procedure consisted of soaking the corncob in 1.0 g/l H_2SO_4 at 60 °C for 12 h, followed by filtration and washing with tap water to pH 6 and steaming at 135-140 °C for 0.5 h. The steamed corncob was mashed and hydrolysed by xylanase [69]. Corn cob has also been utilized for the production of cellulase by *Trichoderma reseii* [70].

To produce XOS with enzymatic methods, xylan is generally extracted with an alkali, such as KOH or NaOH, xylan is converted to XOS by xylanase enzyme having low exo-xylanase and/or β-xylosidase activity [71] (Figure 3). There are many reports describing the production of XOS by enzymatic hydrolysis of xylan, from oat spelt [72, 73], wheat straw [74]. In another study, three commercial xylanase preparations (Rapidase Pomaliq from Gist-Brocades, Clarex ML from Genenor, and Validase from Valley Research) were evaluated as a sole enzyme source for the enzymatic production of pentoses from the hemicellulose fraction of corn husks and corncobs [75]. Products of the enzymatic reaction were identified as arabinose, xylose, xylobiose, and xylotriose. Akpinar et al. [71] demonstrated that cotton stalk, which has no economical value, could be converted by enzymatic hydrolysis to a more valuable XOS product without production of significant quantities of xylose after extraction of its xylan.

A formulation containing both xylanases and *Lactobacillus* was sprayed over silage. Xylanases present in the formulation selectively depolymerized hemilcellulose producing xylose, which was fermented to lactic acid by *Lactobacillus* thereby increasing the stability, digestibility and nutritive values of the silage for digestion by cattle [76, 77].

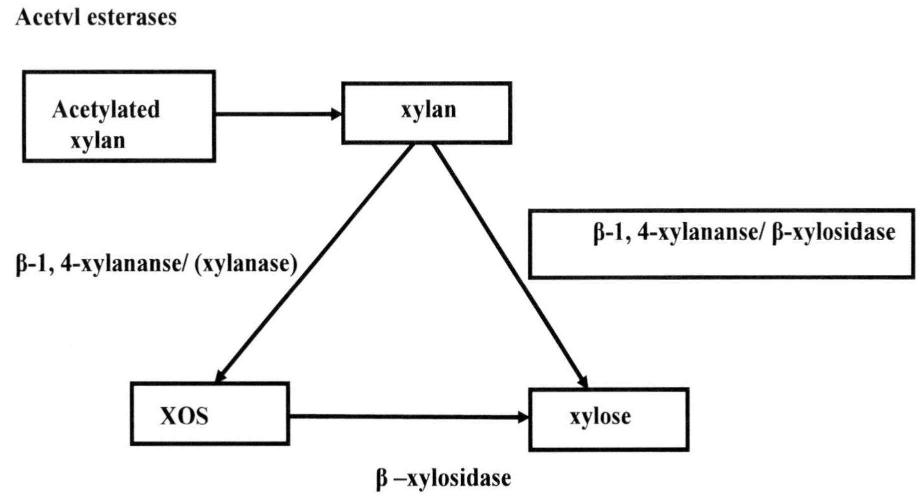

Figure 3. Process flow chart for the production of xylooligosaccharides from lignocellulosic biomass. LCM: lignocellulosic material.

8. By-products of Milling Industry

8.1. Wheat Bran

Wheat bran is the portion of the grain immediately under the husk of the wheat kernel. During the milling process of the wheat grains for the production of wheat flour, large quantities of wheat bran is produced as a by-product. Among the various agro-industrial residues, wheat bran however holds the key, as it is produced worldwide in enormous quantities as an important by-product of the cereal industry. Wheat bran is composed of 8.12% fibre, 15.7% protein, 65% nitrogen, 4.57% ash and 16.7% total sugars. Wheat bran serves as one of the most suitable substrates for production of lignocellulolytic enzymes and pectinases. Pectinolytic enzyme production using wheat bran has been achieved using various microbial cultures such as *Aspergillus niger*, *Aspergillus carbonarious*, *Streptomyces* spp, *Thermoascus auriantacus* [78,79]. Wheat bran has also been used for the production of glucoamylases [80]. Sodhi et al. [81] determined the production of thermostable amylases from *Bacillus* species on wheat bran supplemented with glycerol (1.0%, w/w), soybean meal (1.0%, w/w), L-proline (0.1%, w/w), vitamin B-complex (0.01%) and moistened with tap water containing 1% Tween-40. Xylose from wheat bran was obtained after treatment with β-xylosidase obtained from *Bacillus halodurans* [82]. Ferulic acid is the main phenolic acid in wheat bran which is esterified to arabinose residues in the cell wall arabinoxylans and obtained after enzymatic hydrolysis of wheat bran with ferulolyl esterase [83]. Biovanilin was obtained from ferulic acid [83]. Wheat bran insoluble dietary fibre was treated with xylanases from *Bacillus subtilis* which resulted in release of probiotic ferulolyted oligosaccharides [84]. Xylo-oligosaccharides have also been prepared from wheat bran by treatment with commercial xylanase [85].

8.2. Rice Bran

Rice bran is a by-product of the rice milling process, and it contains various antioxidants that impart beneficial effects to human health. The major rice bran fractions contain 12-13% oil, 11.3-15 % protein, 34-62 % carbohydrates [86]. These components exist in compact form in the rice cell such as protein bodies, amyloplasts and lipid bodies [86]. The use of enzymes in rice bran processing is still a new and relatively unexplored technology. Application of enzymes in rice bran can facilitate the release of valuable cell wall bound constituents. Oil from rice bran has been obtained after enzymatic treatment with proteases. Jahani et al. [87] optimized the processing conditions through response surface methodology for enzymatic degumming of rice ban oil. Rice bran has also been used as a substrate for xylanase production [88]. Protein from heat stabilized rice bran was extracted after amylase pretreatment [89]. In a study, enzyme assisted extraction of nutraceuticals from rice bran was done and further used to control the lipid oxidation in emulsions [90]. Some of the value-added products developed from agricultural wastes are presented in Table 3.

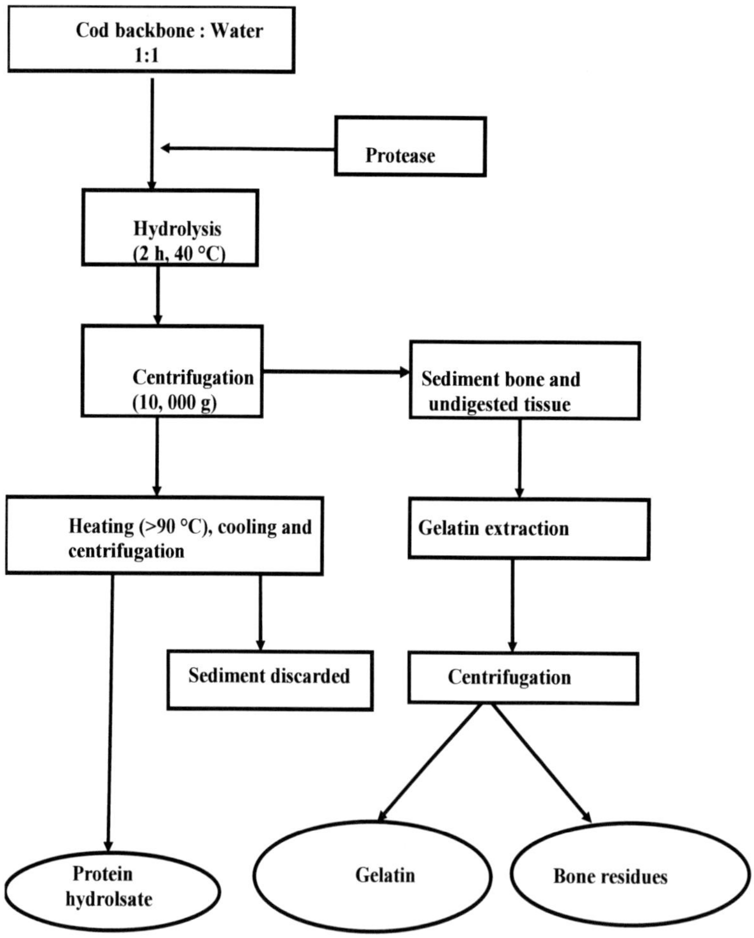

Figure 4. Process diagram for bioconversion of fish waste to value- added products.

Table 3. Value added products from agro-processing wastes

Substrate	Product	Enzyme	References
Orange peel; Kinnow peel	Ethanol	Cellulases and pectinases	[25, 113]
Tomato waste	Vitamin $B_{12;}$ lycopene	Cellulases and pectinases	[114, 115]
Grape pomace	Extraction of phenolic compounds	Pectinases	[41]
Sugarcane bagasse	Citric acid	Cellulases and pectinases	[60]
Flax	Lignans	Cellulase and beta-glucosidase	[116]
Olive oil processing waste	Biodiesel	Lipases	[117]

9. Seafood and Meat Industry Waste

Slaughterhouses as well as the processing of meats generate a significant amount of solid and liquid by-products. Wastes generated by the meat industry are mostly bones, organs and hard tissues containing animal proteins of poor biodegradability. The majority of hard-to-degrade proteins are extracellular matrix proteins (collagen, elastin, proteoglycans) and keratins. Keratin-rich wastes including feathers, hair, nails, horns, etc. are typical by-products of poultry slaughtering and the leather industry. Currently, incineration is the main route for the disposal of such wastes. Some of the value-added products using enzyme technology are: biodiesel obtained after enzymatic transesterification of discarded tallow [91], tallow has also been used for the production of lipase [92], generation of crude protein extracts from keratin through enzymatic hydrolysis with keratinase [93], flavour ingredients containing small peptides and free amino acids with umami taste were produced after hydrolysis of chicken waste proteins with proteases and bromelin [94].

Seafood processing operations generate large quantities of inedible by-products. Seafood industry effluent streams are usually discarded directly into the sea without any treatment or valorization. Discards from sea food processing plants have been reported to contain various biomolecules, such as valuable enzymes, pigments like carotenoids, collagen, gelatin, bioactive peptides etc. Fish visceral discard contains high quantities of digestive enzymes, such as collagenases, trypsin, chymotrypsin, elastase etc. These enzymes can be recovered by suitable methods. Fish protein hydrolysate can be prepared from by-products of fish. Hydrolysis of fish waste proteins can be achieved by both digestive enzymes of fish itself or by addition of external sources of enzymes [95] (Figure 4). Oligomers of chitin and chitosan which are bioactive compounds have also been prepared by enzymatic hydrolysis with proteases [96].

Conclusion

Food processing industry generates a large quantity of waste in the form of straw, stalk, peels, seeds, stone, skin, scales, bones, etc. Most of these wastes are not put to any significant commercial use and are hence disposed off into the municipal bins or in open spaces causing environmental pollution. Though termed as wastes, such non-edible processing output has tremendous potential for further processing owing to their nutritional composition. The striking characteristic of these wastes is that they can act as base material for production of enzymes and the enzymes produced, thereof can be used for bioconversion of wastes into commercially important products. Enzymes, such as cellulases, pectinases, proteases, amylases, xylanases play a substantial role in bioconversion of such wastes into value-added products. As described in this chapter, products like sugars, concentrated sugar syrups, ethanol, organic acids, flavouring compounds, oligosaccharides, etc could be produced from wastes, directly by enzymes or by processes involving enzymes. Although, the processes for production of value-added compounds from food processing wastes have been standardized, some technological refinement, scale-up studies, techno-economic evaluation need to be carried out before these processes are put to any commercial use.

REFERENCES

[1] D. Miljkovic. and G.S. Bignami, Google Patent No 10/992.502.(2002).
[2] L. Wati, S. Kumari and B.S. Kundu, *Ind J Microbiol,* 147, 26(2007).
[3] D.E. Barreto, T.J. Menezes, J.G. Salva, V.L. Baldin, R.S. Papini, A.M. Sales, *Process Biochem,* 24, 167, (1989).
[4] Z. Zheng and K. Shetty, *Process Biochem,* 33: 323, (1998).
[5] Z. Zheng and K. Shetty, *Process Biochem,* 35: 825, (2000).
[6] V.S. Bisaria and T.K. Ghose, *Enz and Microbial Technol* 3: 90–104. (1981).
[7] C.I.N. Lewis and E Yamamoto, *Annual Review of Plant Physiology and Plant Molecular Biology,* 41: 455, (1990).
[8] J. N. Miller, An introduction to pectins: Structure and properties. In: Fishman, M.L., Jem, J.J. (Eds.), Chemistry and Functions of Pectins, ACS Symposium Series 310. *American Chemical Society,* Washington, DC. (1986).
[9] J. Ikram-ul-haq, M.M. Javed, T.S. Khan, *Afr J Biotechnol,* 5: 609, (2006).
[10] S. Ahmed, S. Riaz, A. Jamil, *Appl Microbiol Biotechnol,* 84:19, (2009).
[11] A. Knob, C.R.F. Terrasan, E.C. Carmona, *World J Microbiol and Biotechnol,* 26: 389 (2010).
[12] J.M. Palma, L.M. Sandalio, F.J. Corpas, M.C. Romero-Puertas, I. McCarthy, L.A. Del Rio, *Plant Physiol Biochem,* 40: 521 (2002).
[13] A. Pandey, S. Benjamin, C.R. Soccol, P. Nigam, N. Krieger, V.T. Soccol, *Biotechnol Applied Biochem,* 29: 119 (1999).
[14] K.J. Dussan, C.A. Cardona, O.H. Giraldo, L.F Gutierrez, V.H Perez,. *Bioresour Technol,* 101: 9542 (2010).
[15] S. Kumar, N. Katiyar, P. Ingle, S. Negi, *Bioresour Technol,* 102: 4909 (2010).
[16] A. Pandey, *Process Biochemistry*, 27:, 109 (1992).
[17] A. Pandey, C.R. Soccol, P Nigam, V.T. Soccol, *Bioresour Technol,* 74: 69 (2000).
[18] R. Auria, J. Palacios, S. Revah, Biotechnol Bioeng, 39: 898 (1992).
[19] S Ramachandran, A.K. Patel, K.M. Nampoothiri, F. Francis, V. Nagy, G Szakacs, *Bioresour Technol,* 93:169 (2004).

[20] S. Gorinstein, S. Poovarodom, H. Leontowicz, M. Leontowicz, J. Namiesnik, S. Vearasilp, R. Haruenkit, P. Ruamsuke, E. Katrich, Z. Tashma, Food Res Int, 44: 2222` (2011).
[21] FAOSTAT. (2010) Food and Agriculture Organization, Rome, Italy. www.fao.org.
[22] L. Izquirdo, J.M. Sendra, Citrus fruits composition and characterization . In: Encyclopedia of Food Science and Nutrition, Caballero, B.L Trugo and P. Finglas (Eds.).(pp. 6000). Oxford: Academic Press 1335-1341. (2003)
[23] F.R. Marin, C.R. Solr, G.O. Benavente, J. Castillo, H. Perz, *Food Chem,* 100: 736 (2007)
[24] E.S. Martins, D. Silva, R. Da Silva, E. Gomes, *Process Biochem* 37: 949 (2002).
[25] H.S. Oberoi, Y. Chavan, S. Bansal, G.S. Dhillon, *Food Bioprocess Technol* 3: 528 (2010).
[26] H.S. Oberoi, P.V. Vadlani, A. Nanjundaswam, S. Bansal, S. Singh, S. Kaur, N Babbar, *Bioresour Technol* 102: 1593 (2011).
[27] A. De-Gregorio, G. Mandalari, N. Arena, F Nucita, M.M. Tripodo, R.B. Lo Curto,. *Bioresour Technol,* 83: 89 (2002).
[28] M.M Tripodo, F Lanuzza, G. Micali, R. Coppolino, F. Nucita, *Bioresour Technol,* 91: 111 (2004)
[29] Z. Sroska, W. Cisowski, *Food Chem Toxicol*, 41: 753 (2003).
[30] M. Puri, A. Kaur, W.H. Schwarz, S. Singh, J.F. Kennedy, *Int J Biol Macromoles*, 48: 58 (2011).
[31] S. Bhushan, K. Kalia, M. Sharma, B. Singh, P.S. Ahuja, *Crit Rev Biotechnol*, 28: 285 (2008).
[32] J. Sun, X. Hu, G. Zhao, J. Wu, Z. Wang, F Chen, X. Liao, *Food Sci Technol Int*, 13: 91(2007).
[33] T. Vasilev, I. Yu, A.N. Morozov, G.M. Zaiko, V.G. Moiseeva, *Konservnaya-i-Ovoshchesushil' naya- Promyshlennost,* 3: 32 (1976).
[34] P. Cicro, A. Nogueira, N.C. Smith, G. Wosiacki, *Agrotec*, 29: 40 (2005).
[35] S.G. Vilas-Boas, E. Esposito, M.M Mendonca, *World J Microbiol Biotechnol* 18 : 541(2002).
[36] M.R. Wilkins, W.W. Wilbur, K. Grohmann, R.G. Cameron, *Bioresour Technol* 98: 1596 (2007).
[37] S. Ketnawa, P. Chaiwut, S. Rawdkuen, *Food Sci Technol*, 17: 395 (2011)
[38] C. Krishna, M. Chandraskaran, *Applied Microbiol Biotechnol*, 46: 106.
[39] H.S. Oberoi, P.V. Vadlani, L. Saida, S. Bansal, J. D. Hughes, *Waste Management,* 31: 1576 (2011).
[40] H.S. Oberoi, S.K. Sandhu, P.V. Vadlani, *Food Bioprod Processing,* 90: 257 (2012).
[41] T. Maier, A. Goppert, D.R. Kammerer, A. Schieber, R. Carle, *Eur Food Res Technol,* 227: 267(2008).
[42] S.K. Sharma, L. Maguer, *Italian J Food Sci,* 2: 107 (1996).
[43] N. L. Rozzi, R.K. Singh, R.A. Vierling, B.A. Watkins, *J Agric Food Chem,* 50: 2638 (2002).
[44] E.H. Papaioannou, A.J. Karabelas, *Acta Biochimica Polonica*, 59: 71 (2012).
[45] A.V.L. Pizarro, E.Y. Park, *Process Biochem*, 38: 1077 (2003).
[46] N. W. Li, M.H. Zong, H. Wu, *Process Biochem,* 44: 685 (2009).

[47] N. Dizge, C. Aydiner, D.Y. Imer, M. Bayamoglu, A. Tanriseven, *Bioresour Technol* 100: 1983 (2009).
[48] K.J. Cheng, L.B. Selinger, J.H. Liu, Y. Hu, C.W. Forsberg, M.M. Moloney, *Xylanase obtained from an anaerobic fungus.* US Patent No. 6137032 (2000).
[49] K.A. Clarkson, A.J. Morgan, Z.C. Wang, Xylanase from *Acidothermus cellulolyticus.* US Patent No 5902581(1999).
[50] L.M.F. Gottschalk, R.A. Oliveira, E.P.D.S. Bon, *Biochem Eng J,* 51: 72 (2010).
[51] V. Micard, C.M.G.C Renard, R.J.F. Thibault, *Enz Microbial Technol,* 19: 162 (1996).
[52] J.C. Olmos, M.E.Z. Hansen, *Chem Eng J,* 192: 29 (2012).
[53] F. Michel, J.F. Thibault, J.L. Barry, R. D. Baynast, *J Sci Food Agric*, 2: 77 (1988).
[54] M. Spagnuodo, C. Crecchio, M.D.R Pizzigallo, P. Ruggieo, *Bioresour Technol*, 60: 215 (1997).
[55] G. Schaffeld, A. Illanes, G. Mejias, A.M. Pinochet, *J Chem Technol Biotechnol*, 39: 161(1987).
[56] I. E. Baciu, H.J. Jordening, *Enz Microbial Technol*, 34: 505 (2004).
[57] M.D. Sutton, J.B.D. Peterson, *J Sugar beet Res*, 38: 19 (2001).
[58] Z.H. Bai, H.X. Zhang, H.Y. Qi, X.W. Peng, B.J. Li, *Bioresour Technol*, 95: 49 (2004).
[59] M. Pal, A.M. Calvo, M.C. Terron, A.E. Gonzalez, *World J Microbiol Biotechnol*, 11: 541(1995).
[60] H.K. Manonmani, K.B. Srikantiah, *J Food Sci Technol*, 25: 159 (1988).
[61] P. Kumar, T. Satyanarayana, *Lett Applied Microbiol*, 45: 392 (2007).
[62] S. Rodriguez-Couto, M.A. Sanroman, *Biochem Eng J*, 22: 211 (2005).
[63] Y.P. Zhang, L.R. Lynd, *Biotechnol Bioeng*, 88: 797 (2004).
[64] M.I. Massadeh, Wan-Yousoff, W.M. O. Omar, J. Kader, *Lett Biotechnol*, 23: 1771 (2001).
[65] K.N. Niladevi, P. Prema, *Actinomycetol*, 19: 40 (2005).
[66] M.S. Chandra, B. Viswanath, B.R. Reddy, *Ind J Microbiol,* 47: 323 (2007).
[67] B.S. Chadha, S.S. Kanwar, H. S. Garcha, *Actamicrobiol Immuno Hung*. 42: 71 (1995).
[68] P.T. Sangeetha, M.N. Ramesh, S.G. Prapulla, *Applied Microbiol Biotechnol,* 65: 530 (2004).
[69] R. Yang, S. Xu, Z. Wang, W. Yang, *LWT Food Sci Technol*, 38: 677 (2003).
[70] X. Liming, S. Xueling, *Bioresour Technol*, 91: 259 (2004).
[71] O. Akpinar, O. Ak, A. Kavas, U. Bakir, L. Yilmaz, *J Agric Food Chem*, 55: 5544 (2007).
[72] C. Chen, J. Chen, T. Lin, *Enz Microbial Technol*, 21: 91 (1997).
[73] A.A. Aachary, S.G. Prapulla, *Bioresour Technol*, 100: 991 (2009).
[74] C. Zilliox, P. Debeire, *Enz Microbial Technol*, 22: 58 (1998).
[75] K.Y. Yoon, M. Cha, S.R, Shin, K.S. Kim, *Food Chem*, 92, 151(2006).
[76] C.T. Evans, S.P. Mann, R.C. Charley, D. Parfitt, US Patent 5432074 (1995).
[77] C.T. Evans, S.P. Mann, R.C. Charley, D. Parfitt, Formulation for treating silage. *European Patent* 0563133 (1995b).
[78] M. Kapoor, R.C Kuhad, *Letts Applied Microbiol* 34: 317 (2002).
[79] R.C. Kuhad, M. Kappor, R. Rustagi, *World J Microbiol Biotechnol*, 20: 257 (2004).
[80] A. Pandey, *Biological Wastes,* 34: 11 (1990).
[81] H. K. Sodhi, K. Sharma, J.K. Gupta, S.K. Soni, *Process Biochem,* 40: 525 (2005).
[82] I. Smaali, I. Remond, Y. Skhiri, M. J. O'Donohue, *Bioresour Technol,* 100: 338 (2009).

[83] D.D. Gioia, L. Sciubba, L. Setti, F. Uziatelli, M. Ruzzi, D. Zanichelli, F. Fava, *Enz Microbial Technol*, 41: 498 (2007).
[84] H. Yuan, J. Wang, H. Yao, *Food Chem*, 90: 759 (2005).
[85] C. Manisseri, M. Gudipati, *LWT Food Sci Technol*, 43: 421 (2010).
[86] B.O. Juliano, Rice in human nutrition, first ed. International rice research Institute/FAO *Rome* pp 35-45. (1993).
[87] M. Jahani, M. Alizadeh, M. Pirozifard, A. Qudsevali, *LWT Food Sci Technol*, 41: 1892 (2008).
[88] S.L. Wang, Y.H. Yen, I.L. Shih, A.C. Chang, W.T. Chang, W.C. Wu, Y.D. Chai, Enz *Microbial Technol,* 33: 917 (2003).
[89] N. Xia, J. Wang, X. Yang, S. Yin, J. Qi, L. Hu, X. Zhou, *J Food Eng*, 110: 95 (2012).
[90] E. Revilla, C.S. Maria, E. Miramontes, J. Bautista, A.G. Martinez, O. Cremades, R Cert, J. Parrado, *Food Res Int*, 387: 393 (2009).
[91] H.N. Bahtti, M.A. Hanif, M. Qasim, A.U. Rheman, *Fuel* 87: 2961(2008).
[92] K. Ramani, L.J. Kennedy, M. Ramakrishnan, G. Sekaran, *Process Biochem,* 45: 1683 (2010).
[93] B. Deivasigamani, K.M. Alagappan, *J Environ Biol*, 29: 933 (2008).
[94] K. Maehashi, M. Matsuzaki, Y. Yamamoto, S. Udaka, *Biosci Biotechnol Biochem*, 63: 555 (1999).
[95] S.K. Kim, E. Mendis, *Food Res Int*, 39: 383 (2006).
[96] M. Ropars, R. Marchal, A.J. Pourquie, J.P. Vandecastelee, *Bioresour Technol,* 42: 197 (1992).
[97] A.B. Bjerre, A.B. Olesen, T. Fernqvist, A. Ploeger, A.S. Schmidt, *Biotechnol Bioeng*, 49: 568 (1996).
[98] H.S. Oberoi, N. Babbar, S.K. Sandhu, S.S. Dhaliwal, U Kaur, B.S. Chadha, V.K. Bhargav, *J Ind Microbiol Biotechno,l* 39: 557 (2012).
[99] R.F.H. Dekker, A.F.A. Wallis, *Biotechnol Bioeng*, 25: 3027 (1983).
[100] M.C Maldonado, A.M.S. Saad, *J Ind Microbiol Biotechnol*, 20:34 (1998).
[101] D. Madamwar, S. Patel, H. Parik, *J Ferment Bioeng,* 67: 424 (1989).
[102] M.S. Revankar, K.M. Desai, S.S. Lele, *App Biochem Biotechnol,* 143: 16 (2007).
[103] S.A, Singh, M. Ramakrishna, A.G.A. Rao, *Process Biochem* 35: 411 (1999).
[104] T. Vares, M. Kalsi, A. Hatakka, *Appl Environ Microbio,l* 61: 3515 (1995).
[105] E. Rosales, S. Rodriguez Couto, M.A. Sanroman, *J Food Eng,* 66: 419 (2005)
[106] M.J.F. Vieira, A.C.C. Spadaro, S. Said, *Lett Biotechnol*, 13: 39 (1991).
[107] P.Nigam, A. Pandey, K. A. Prabhu, *Biological Wastes,* 19: 275 (1987).
[108] J.F. Osma, J.L. Toca Herrera, S. Rodriguez Couto, *Dyes Pigments,* 75: 32 (2007).
[109] C. Botella, A. Diaz, I. Ory, C. Webb, A. Blandino, *Process Biochem*, 42: 98 (2007).
[110] B.C. Saha, *Bioresour Technol*, 90: 33(2003).
[111] H.S. Oberoi, S.K. Kalra, D.S. Uppal, S.K. Tyagi, *Int J Food Sci Technol*, 42: 228 (2007).
[112] K. Brijwani, H.S. Oberoi, P.V. Vadlani, *Process Biochem,* 45: 120 (2010)
[113] S.K. Sandhu, H.S. Oberoi, S.S. Dhaliwal, N. Babbar, U. Kaur, D. Nanda, D. Kumar, *Annals of Microbiol,* 62: 655 (2012).
[114] M.S.Y. Haddadin, I.M.A. Reesh, F.A.S. Haddadin, R.K Obinson, *Bioresour Technol,* 78: 225 (2001).
[115] A. Zuorro, M. Fidaleo, R. Lavecchia, *Enz Microbial Technol*, 49: 567 (2011).

[116] S. Renouard, C. Hano, C. Corbin, O. Flinlaux, T. Lopez, J. Montguillon, E. Barakzoy, F. Mesnard, F. Lamblin, E. Laine, *Food Chem,* 122: 679 (2010).
[117] C.F. Torres, A.M. Tore, T. Fornari, F.J. Seorans, G. Reglero, *Biochem Eng J*, 34: 165 (2007).

In: Enzymes in Value-Addition of Wastes
Editors: Satinder Kaur Brar and Mausam Verma
ISBN: 978-1-62948-968-1

Chapter 3

SOLID STATE FERMENTATION OF AGRICULTURAL WASTES FOR THE PRODUCTION OF BIOCATALYSTS

A. Ganesh Kumar*, S. Senthil Kumar, G. Dharani and R. Kirubagaran
Marine Biotechnology Division,
Earth System Sciences Organization - National Institute of Ocean Technology (ESSO-NIOT),
(Ministry of Earth Sciences, Govt. of India)
Pallikaranai, Chennai, Tamil Nadu, India

ABSTRACT

Agricultural wastes contain several reusable substances of high value which are not utilized or underutilized. Disposal of agricultural wastes directly by thermal incineration, land filling, or by other improper handling has severe impact provoking detrimental effect on ecosystems and the environment. These lignin and cellulose rich wastes have immense potential in production of enzyme where they can supply and become economical source of raw materials. The production of enzymes from these wastes by solid-state fermentation (SSF) process is a viable alternative and eco-friendly process. This process can be carried out using low chemical and energy input. Solid wastes can be used as support material or as a source of potent nutrients in SSF to produce enzymes with overwhelming economical benefits. Furthermore, the enzymes obtained through SSF are of superior quality and can be used as an appropriate biocatalyst for multitude functions. SSF fetches two major functions, utilization of solid wastes as substrates and production of value added products. This chapter gives an insight on various supports used, fungal and bacterial enzymes production, fermentation conditions of SSF and strain improvement strategies for optimized production of enzymes.

Keywords: Enzymes, solid state fermentation, non-inert support, inert support, strain improvement

* Correspondence: Dr. A. Ganesh Kumar. Phone: +91 - 44 - 6678 3424 ; Fax: +91 - 44 - 6678 3430 ; E-mail: microganesh@yahoo.com; microganesh@niot.res.in.

8.1. Introduction

8.1.1. Enzymes

Enzymes are natural biocatalysts which have significant biological, catalytic and physicochemical properties. Enhanced rate of reaction and specificity renders microbial enzymes highly perspective for various industrial applications. Nowadays, enzymes are being systematically developed as eco-friendly catalysts for wide ranges of commercial applications (Zhang and Kim, 2010). Enzymes have major application in the fields of industrial, food, medical and environmental applications. In the forthcoming years it is presumed that microbial enzymes will considerably increase their prevalence in the world market. Enzymes as biocatalysts have remarkable advantages over chemical or live-cell based catalysts for commercial applications. The leading part of the enzymes market is occupied by hydrolases (proteases, carbohydrases and esterases). In contradiction to the high existence of oxidoreductases in nature, laccases are some of the few oxidoreductases that are commercialized widely (Kirk et al. 2002).

8.1.2. Enzyme Production

Microbial enzyme production is an economically viable area in the emerging field of biotechnology. Mostly enzyme manufacturers produce enzymes by submerged fermentation (SmF) techniques. However, SSF holds tremendous potential for the production of enzymes with high activity. SSF comprise growth and metabolism of the microorganisms in absence of any free flowing water. SSF mimics the natural habitats of certain microbes and thus it is more efficacious in producing high levels of enzymes and metabolites. In SSF, microbial growth and product formation take place on the surfaces of solid substrates. It has immense advantage over SmF including usage of inexpensive media, cost-effective and ecofriendly downstream processing, less contamination, influent usage and effluent generation (Pandey et al. 2001).

8.1.3. SSF Substrates

Inexpensive agricultural wastes represent one of the most energy rich sources due to its high carbon reserves and are widely considered as an excellent substrates for SSF operation. In recent years, there has been an extensive progress towards more efficient utilization of agricultural residues for the cultivation of microorganisms to produce various enzymes. SSF is generally represented by the growth of microorganisms, on and/or within particles of a solid substrate in the presence of differing moisture levels. The solid substrate serves as a source of carbon, nitrogen, minerals and vital growth factors, and has a capacity to absorb water essential for microbial growth. Some of the agricultural wastes that are being used as SSF substrates include bran, bagasse, straw, husk, hull, trimmings, coir, pith, pulp, etc. These residues are classified into two major groups based on nutritional properties: fibrous residues (low and high digestibility) and brans. Fibrous residues are divided into those with high digestibility, *e.g.*, soy and peanut husk, and those with low digestibility, *e.g.*, sugarcane bagasse and cotton straw. Second major group of agriculture residues is brans, which includes wheat, rice, corn, barley, millet and peanut bran (Graminha et al. 2008). In SSF, the solid substrates not only provide nutrients to the culture, but also shelter and support the growth of microorganisms (Baysal et al. 2003). Considering the cost and availability, the selection of an appropriate solid substrate plays a predominant role in the development of efficient SSF processes (Elibol and Moreira, 2005). The significance of SSF comes from its simplicity and proximity to the natural growth conditions of many microorganisms.

8.2. Nature of Solid Support on SSF

The nature of solid substrate is a vital factor influencing SSF, thus microorganisms produce specific enzymes and metabolites than any other fermentation methods (Pandey, 2003). Solid substrate used in SSF addresses two major functions: supply of the nutrients to the culture and serving as an anchorage for the microbial growth. The materials used in SSF can be divided into two categories: non-inert (organic) and inert (synthetic) materials. Inert materials are used only for attachment of cultures, whereas non-inert functions both as support matrix and as a source of nutrients. The rate of production and quality of the enzyme secreted depends much on the size and active ingredients of the substrates (Hesseltine, 1987).

In case of non-inert materials, a divided and humidified solid (e.g. cereal grain, flour, bran and sawdust) serves as support as well as source of micro and macro nutrients, due to which it is termed as support-substrate. Utilization of agricultural wastes as support-substrate in SSF processes provides an alternative strategy in value-addition to the wastes. In the inert material group, a solid matrix which is nutritionally inert for example polyurethane foam (PUF) is soaked in a nutrient solution acts exclusively as a support. The use of synthetic material to produce enzyme is less pursued and shows certain economical disadvantages. Even though this method is not suitable, it has other uses such a specified medium components and homogenous inert support materials controlling the fermentation process. Interestingly, both the SSF processes have shown to be more suitable for the production of enzymes by filamentous fungi, since they imitate their natural environmental conditions.

8.2.1. Enzyme Production – Non-Inert (natural) Support

Agricultural wastes are one of the ideal substrates for cultivation of microbial culture and optimized enzyme production in SSF. These materials are typically starch or lignocellulose-based agricultural products or agro-industrial residues such as cassava, potato, soybean, sugar beet, sweet potato, and sweet sorghum, crop residues such as bran and straw of rice and wheat, bran of maize, ragi, green gram, black gram, red gram, corn and rice, bagasse of sugarcane and cassava, residues of the coffee processing industry such as coffee pulp and coffee husk, residues of fruit-processing industries such as pomace of apple and grapes, waste of oil-processing mills such as coconut cake, soybean cake, peanut cake, canola meal and palm oil mill waste, and others such as saw-dust, corn cobs and chicory roots, etc. Table 1 shows the production of various enzymes using different natural support systems through SSF technique.

Table 1. Examples of SSF using natural solid supports

Enzymes	Species	SSF Substrate	Reference
Acid protease	*M. bacilliformis*	Wheat bran	Fernandez-Lahore et al. 1998
Alkaline protease	*B. altitudinis* GVC11	Castor husk	Madhuri et al. 2012
α-amylase	*B. subtilis* CBTK 106	Banana waste	Krishna and Chandrasekaran, 1996
α-L-arabinofuranosidase	*Arthrobacter* sp. MTCC 5214	Wheat bran and oat bran	Khandeparker et al. 2008
L-asparaginase	*Cladosporium sp.*	Wheat bran	Mohan Kumar et al. 2013
Cellulase	*A. terreus*	Rice straw	Narra et al. 2012
Chitinase	*O. xanthineolytica* NCIM 2839	Wheat bran	Waghmare et al. 2011
Endo-1,4-β-D-glucanase	*B. subtilis* KIBGE HAS	Sugarcane bagasse	Bano et al. 2013
Endo-β-1,4-xylanase	*A. fumigatus* FBSPE-05	Wheat bran, Sugarcane bagasse, brewer's spent grain	Souza et al. 2012
Feruloyl esterase	*A. niger* I-1472	Sugar beet pulp	Asther et al. 2002
Fructosyl transferase	*A. oryzae* CFR 202	Cereal bran, corn products	Sangeetha et al. 2004
α-galactosidase	*S. griseoloalbus*	Soya bean flour	Anisha et al. 2008
Glucoamylase	*Aspergillus sp.HA-2*	Rice flake manufacturing wastes and wheat bran	Anto et al. 2006
Glucose oxidase	*A. niger* NRRL 3	Buckwheat seed	Ramachandran et al. 2008
β-glucosidase	*H. brevis var. thermoidea*	Wheat bran	Masui et al. 2012
Glutaminase	*Z. rouxii*	Oatmeal and wheat bran	Iyer and Singhal, 2010
	Z. rouxii NRRL-Y 2547	Wheat bran and sesamum oil cake	Kashyap et al. 2002
Inulinase	*K. marxianus* NRRL Y-7571	Sugarcane bagasse	Mazutti et al. 2006
Invertase	*Fusarium* sp.	Wheat bran	Shaheen et al. 2008
Laccase	*P. ostreatus*	Sugarcane bagasse	Karp et al. 2012
Lignin peroxidise (LiP)	*T. versicolor* IBL-04.	Corncobs	Asgher et al. 2012
LiP and manganese peroxidase (MnP)	*P. chrysosporium*	Sapota leaves	Ganesh Kumar et al. 2006
Lichenase	*B. licheniformis* UEB CF	Pea pomace	Chaari et al. 2012
Lipase	*R. homothallicus*	Sugarcane bagasse	Rodriguez et al. 2006
MnP	*L. edodes*	Eucalyptus residue	Arantes et al. 2011
Pectinase	*A. niger* IMI 91881 and *A. sojae* ATCC 20235	Wheat bran and dried orange peel	Heerd et al. 2012
Phytase	*A. ficuum*	Lentils	Bennett and Yang, 2012
Polygalacturonase	*A. sojae* ATCC 20235	Crushed maize	Ustok et al. 2007
Protease	*P. aeruginosa* PseA	Jatropha curcas seed cake	Mahanta et al. 2008
Synthetic lipase	*R. chinensis*	Wheat flour and wheat bran	Sun et al. 2008
Tannase	*P. variotii*	Orange pomace	Madeira et al. 2012
Xylanase	*A. terreus* MTCC 8661	Palm fiber	Lakshmi et al. 2009

8.2.2. Enzyme production – Inert Support

Studying fermentation variables in the conventional SSF systems, where substrates are mostly agro-industrial products and sub-products such as wheat and rice bran, castor husk, oatmeal, soybean flour, sugarcane bagasse, fruit peels, etc, presents certain drawbacks, such as difficulty in process control and presence of impurities in the end product. Moreover, the possibilities of contamination is much high which directly affects the purity of the metabolites production. The use of inert support in SSF will play some important role in providing a clean and uncontaminated environment for the production of the enzymes, along with good environment for controlling and monitoring microbial metabolism (Auria, 1990). Table 2 presents examples of the wide range of applications of SSF on enzyme production using inert supporting matrix.

Table 2. Examples of solid-state fermentation using impregnated inert supports

Enzymes	Microorganism	Support	References
Amylase	*A. oryzae*	PUF	Murado et al. 1997
Cellulase	*T. veride*	PUF	Hu et al. 2011
Chitinase	*L. lecanii*	PUF	Marin-Cervantes et al. 2008
Glutaminase	*V. costicola*	Polystyrene	Prabhu and Chandrasekaran, 1997
	Beauveria sp.	Polystyrene	Sabu et al. 2000
Glucose oxidase	*A. niger*	PUF	Miron et al. 2010
Invertase	*A. niger*	PUF	Montiel-Gonzalez et al. 2004
Laccase	*T. pubescens*	Nylon sponge, PUF, stainless steel sponges	Rodriguez-Couto, 2012
Nuclease	*P. citrinum*	PUF	Zhu et al. 1996
Phytase	*A. ficuum* and *R. oligosporus*	Polystyrene	Gautam et al. 2002

8.2.3. Selection of the Support

Compared with natural substrates the utilization of inert supports in SSF clearly facilitates specific enzyme synthesis and improves the downstream processing (Prabhu and Chandrasekaran 1995). Inert supports impregnated with defined nutrients will advance to be a valuable research tool, and will become more important for large scale industrial processes, as they exhibit certain advantages that are not obtained with natural substrates. Strikingly, SSF on inert supports facilitates a more-reproducible fermentation process, an important and desired characteristic for any industrial fermentation process. However, economics will be one of the major significant factors in determining the success of SSF on inert supports. Rodriguez-Couto et al. (2002) observed the difference in laccase production profile in SSF using both inert and non inert supports. In PUF support, the laccase activity appeared on the 7th day (28 U/L) and then increased reaching a maximum value of 204 U/L on the 17th day. In non-inert wood shavings support laccase activity started on the 3rd day (184 U/L), and then increased progressively to reach a maximum value of 451 U/L on the 18th day. The enzyme produced is double in natural support than that obtained employing PUF as a support. In non-inert wheat straw support laccase activity initiated on the 3rd day (195 U/L) and reached a maximum value of 490 U/L on the 17th day. The enzyme activity is double forth higher than that obtained using PUF as a support material. In non-inert barley straw support laccase activity appeared on the 3rd day (279 U/L), and then increased progressively reaching a maximum value of 591 U/L on the 18th day. The enzyme activity is three forth higher than that obtained using PUF as a support material. Interestingly, in non-inert barley bran support laccase activity laccase activity began on the 4th day (130 U/L) and then increased linearly attaining a maximum value of 1155 U/L on the 18th day. The enzyme activity is six forth higher than that attained on the PUF supported SSF. These observations clearly substantiate that SSF on non-inert supports led to higher activities than inert PUF, especially when using barley bran as a support material. Desirable outcome may be due to the provision to the fungus a favorable environment that mimics its natural habitat, where the fungus is more stimulated for the secretion of high amount of ligninolytic enzymes. Another reasonable cause may be the cellulosic constituents of these types of support, which could act as an activator of laccase activity (Srinivasan et al. 1995). The non-inert supports provide nutrients favoring the microbial growth and metabolite production, due to which the initial amount of carbon in the culture medium was decreased to 80%, providing a much reduction in process costs. This signifies the importance of utilizing non-inert agricultural residues as support in SSF for developing countries where its availability is profuse.

8.3. SSF and SmF for the Production of Enzymes

Fermentation can be either homo-fermentative – producing single main product, or hetero-fermentative – producing mixture of products. Versatile range of microorganisms, including fungi and bacteria yield an invaluable array of vital enzymes utilizing appropriate substrates. Production of enzymes has been carried out both by SSF and SmF. Most commercial scale enzyme manufacturers produce enzymes by SmF techniques because of the ease of handling, controlling and monitoring of fermentation. The substrates are utilized quite rapidly and the metabolites are secreted into the liquid broth. Moreover, SmF requires constant supplementation of nutrients for its continuous operation. This technique is best suited for microorganisms which require high level of water. In SSF technique, the substrate utilization is slow and steady, providing optimal growth of microorganisms by controlled release of nutrients. Interestingly, in the last few decades there has been an increasing trend towards the use of the SSF techniques to produce variety of commercially valuable enzymes.

8.3.1. Fungal Enzymes

A direct comparison between SSF and SmF cultivation techniques, steps involved and yield coefficient is difficult because the two processes are entirely different. Studies on fungal enzyme production in SSF have shown that, SSF when compared to SmF provides higher productivity, less prone to carbon catabolite inhibition and yields enzymes with high stability

(Holker et al. 2004). The fungi exhibiting enormous metabolic differences in production of enzymes between SSF and SmF are presented in Table 3. Filamentous fungi are considered as the most prominent microorganisms in SSF due to its ability to grow and multiply well on solid materials with low moisture content. Filamentous hyphal branching mode of growth, good tolerance towards lower levels of moisture and high osmotic pressure conditions make fungi efficient and competitive microflora for utilization of solid substrates for value added product production (Pandey et al. 1994). The identification and selection of appropriate strains for SSF depends mainly on factors such as the nature of the substrate and environmental conditions (Nigam and Singh, 1994). Therefore, the selection of a particular strain remains a much valuable task, especially when commercially viable enzymes are to be achieved.

Table 3. Substrate wise comparison of enzyme productivity by employing various fungi

Fungi	Enzymes	Substrate		Productivity		References
		SSF	SmF	SSF	SmF	
A. niger I-1472	Esterase	Sugar beet pulp	Basal media	20 nkat/g	0.4 nkat/mL	Asther et al. 2002
A. niger NRRL 330	ß-fructofurano-sidase	Sugarcane bagasse	Basal media	81.8 U/L	18.3 U/L	Ashok Kumar et al. 2001
A. niger	Pectinesterase	Sugar cane bagasse pith	Basal media and pectin	725 U/L	127 U/L	Maldonado and Strasser de Saad, 1998
A. niger NCIM1207	Lipase	Wheat bran and olive oil	Synthetic oil based Media (*Aspergillus* minimal medium with 1% olive oil)	630 U/g	18 U/mL	Mahadik et al. 2002
A. niger	Pectinases	Deseeded sunflower head	Sunflower head powder (8g) was added to obtain 2% pectin content in the basal medium	34.2 U/g	12.6 U/mL	Patil and Dayanand, 2006
A. niger	Tannase	PUF and media	Basal media and Tannic acid 2% (w/v)	1861.6 U/mL	14.7 U/mL	Renovato et al. 2011
A. niveus	Polygalactu-ronase	Lemon peel	Czapeck medium and Lemon peel	1324 U/mL	300 U/mL	Maller et al. 2011
A. oryzae NRRL 1808	Protease	Wheat bran	Basal media supplemented with various agro-industrial residues	31.2 U/gds	8.7 U/gds	Sandhya et al. 2005
R. chinensis CCTCC M201021	Synthetic lipase	Wheat bran and wheat flour (w/w, 2/3) and olive oil	Minimal medium with olive oil	24.4 U/g	13.8 U/mL	Sun and Xu, 2009
P. janthinellum EU2D-21 (mutant)	Cellulase	Wheat bran (4 g) and tissue paper (1 g)	Wheat bran (2.5%) and tissue paper (1%)	38 U/g	2.8 U/mL	Singhvi et al. 2011
Penicillium sp.	Lipase	Soybean bran	Media with olive oil	15 U/mL	11 U/mL	Rigo et al. 2010
P. viridicatum RFC3	Pectate Lyase	Wheat bran and orange bagasse	Basal media supplemented with wheat bran and orange bagasse	1500 U/mL or 300 U/g	1000–1200 U/mL	Ferreira et al. 2010

nkatal (nkat): 1 nkat is defined as the amount of enzyme that catalyses the release of 1 nmol of cinnamic acid per sec, gds: gram dry substrate, gss: gram solid substrate.

8.3.2. Bacterial Enzymes

Bacteria produces a wide range of biological products using fermentation techniques, for e.g., enzymes like cellulases, asparaginases, amylases, xylanases etc. are produced through SmF. Recently, it is shown that SSF is more efficient than SmF for the production of enzymes. This may be due to the metabolic differences of organisms as well as the media where the enzymes are produced. One of the major disadvantages in case of enzyme production using SmF was the accumulation of a variety of intermediate metabolites resulting in lowered enzyme activity (Table 4).

Table 4. Substrate wise comparison of enzyme productivity by employing various bacteria

Bacteria	Enzyme	Substrate		Productivity		References
		SSF	SmF	SSF	SmF	
Bacillus sp (Thermophilic)	Xylanase	wheat bran supplemented with mineral salts and xylose	wheat bran supplemented with mineral salts and xylose	13.1 U/ml	12.8 U/ml	Gupta and Kar, 2009
Bacillus sp.	Cellulase	Banana waste	Carboxymethyl-cellulose / glycerol	9.6 U/gds	1.2 U/ml	Sukumaran et al. 2005
Bacillus sp.	Amylase	Wheat bran, rice bran	Nutrient broth and glucose (1%)	41.2 mg/mL	4.3 mg/mL	Raj and Hemashenpagam , 2012
B. pumilus	Endo-xylanase	Wheat bran	Wheat bran	35,156 U/ gds	430 U/ mL	Asha Poorna and Prema, 2006
B. subtilis US417	Phytase	Wheat bran	Wheat bran	1.2 U/g	2.3 U/g	Kammoun et al. 2012
Enterobacter sp NRG4	Chitinase	Chitin flakes rice bran and wheat bran	Cell wall of *Candida albicans* and *Fusarium oxysporum*	965 U/g	49.7 U/ml	Dahiya et al. 2005

8.4. Optimization of Fermentation Conditions for Enzyme Production on SSF

SSF has been envisioned by many scientists as an economically viable and eco-friendly strategy for converting agricultural wastes to valuable enzymes using different types of microorganisms. Intrinsic factors such as temperature, pH, moisture, oxygen levels and concentrations of available micro and macro nutrients have direct impact on the microbial growth in SSF which in turn affects the enzyme production. Whereas, in case of SmF the homogeneity of the microbial suspension and nutrients in the liquid phase makes the control of fermentation quite simple. In SSF, the low moisture content makes the down-streaming process very easier; however, much consideration has to be given for mixing of substrates, inter-exchange of heat, oxygen diffusion, moisture level, gradients of pH changes, nutrient limitation and product formation.

8.4.1. Effect of Incubation Time

Critical success factor in enzyme production using SSF is purely dependent on the time duration of microorganism which can withstand at certain time frames beyond which the regulation of enzymes production will be affected due to self-catalytic inactivation nature of enzymes. It is notable in many cases, the time required for the optimum enzyme production may be as long as 48 - 72 h for bacteria and 8–10 days for fungi (Puri et al. 2002; Aikat and Bhattacharyya, 2000). Prakasham et al. (2005) showed alkaline protease production using green gram husk by *B. circulans* at optimized incubation time of 60 h. Mukherjee et al. (2008) demonstrated alkaline protease production by a thermophilic *B. subtilis* under SSF using Imperata cylindrical grass within a short incubation period of 24 h. Yadav et al. (2011) observed maximum serine protease production by *A. flavus* MTCC 9952 in wheat bran and corn cob mixture (1:1) incubated for 48 h. In *A. oryzae* strain, neutral protease production with wheat bran as solid substrate peaked maximum at 72 h and further incubation showed a rapid decline in the enzyme yield (Sandhya et al. 2005). Likewise, the optimized incubation time for protease production by *Penicillium* sp. showed maximum activity after 72 h (Agarwal et al. 2003). However, in case of *Streptomyces* sp. maximum protease production was obtained after 120 h incubation (De Azeredo et al. 2004). Some researchers observed substrate dependent variations in the protease synthesis, Chutmanop et al. (2008) observed higher protease activity in *A. oryzae* after 60 h on both wheat and rice bran substrates, where rice bran yielded considerably high protease than wheat bran. The biomass specific production of proteases was distinctly high on rice bran and the enzyme activity on rice bran was about 1400 U/gds compared with 1000 U/gds obtained on wheat bran. Under optimal conditions the production of exo-polygalacturonase by *P. viridicatum* RFC3, using mixture of wheat bran and orange bagasse (1:1, w/w) showed enzyme production at 144 h and peaked at 336 h of fermentation (Silva et al. 2007). Similar observation was recorded for the production of thermostable pectinases from thermophilic *T. aurantiacus* when cultured on orange bagasse (Martins et al. 2002).

Reddy et al. (2003) illustrated ability of *P. ostreatus* and *P. sajor-caju* to produce various lignolytic and cellulolytic enzymes such as laccase, lignin peroxidase, xylanase, endo and exo-1,4-β- D-glucanase in SSF of banana leaf biomass and pseudostems. In various extracellular enzymes produced by *P. ostreatus* on banana leaf biomass the specific activities of laccase and LiP were increased from 10 to 20th day and decreased gradually. The maximum specific activities of laccase and peroxidase were 1.71 and 0.16 U/mg, and in xylanase, the maximum specific activity was 0.14 U/mg on day 20. In the various extracellular enzymes produced by *P. sajor-caju* on banana leaf biomass the highest specific activity of laccase was observed on day 10 (1.66 U/mg), followed by LiP on day 20 (0.47 U/mg) and xylanase specific activity was 0.14 U/mg on day 10. Banana leaf biomass was found to be more suitable substrate compared to pseudostem for enzyme production. Papinutti and Forchiassin (2007) investigated the production of lignocellulosic enzymes from *F. sclerodermeus* using soy and wheat bran. The lignocellulosic enzymes cellulases, xylanases and pectinases in soy bran medium reached the optimum production at fermentation time of 12 days, while maximal MnP (14.5 U/g) and laccase (520 U/g) activities were observed at 15 and 28 days, respectively. Ganesh Kumar et al. (2006) observed maximal LiP (2100 U/L) and MnP (1200 U/L) on 7th day SSF of *A. zapota* leaves by *P. chrysosporium* and both activities declined gradually with the progress of fermentation. Many researchers have attempted to optimize the production of enzyme tannase. The production of enzyme tannase by SSF using *R. oryzae* with wheat bran soaked in tannic acid showed maximum production on 4th day of incubation period (Chatterjee et al. 1996). In SSF with jamun leaves using *A. ruber*, the maximum production of tannase was observed after 96 h of incubation (Kumar et al. 2007). Maximum tannase production (0.85 U/gds) by *A. niger* ATCC 16620 was obtained at 72 h when SSF was carried out using coffee husk supplemented with 0.6% tannic acid (Sabu et al. 2005).

8.4.2. Effect of Catabolite Repression

Gessesse and Mamo (1999) reported supplement of different sugars to wheat bran resulted in mild to strong repression of xylanase production. In particular, xylose strongly repressed xylanase production at very low concentration, whereas catabolite repression by glucose was found to be dependent on its concentration, mild at lower concentration (5% w/w) and much distinct at concentrations of 10% and 15% (w/w). Nandakumar et al. (1999) observed negligible catabolite repression by glucose even at higher concentration (150 mg/g wheat bran) on α-amylase and amyloglucosidase production in *A. niger* CFTRI 1105.

Interestingly, Yadav et al. (2011) found addition of external carbon sources galactose, glucose and sorbitol to SSF of wheat bran and corn cob mixture (1:1) enhanced the serine protease to 1949, 2054 and 2277 U/g respectively by *A. flavus* MTCC 9952. However, the higher concentration of sorbitol exhibit inhibitory effect on enzyme yield which may be due to catabolite repression. Certain study also reveals that the effect of carbon catabolite repression is minimal in SSF. Kamra and Satyanarayana (2004) demonstrated xylanase production by *H. lanuginosa* was unaffected by addition of xylose upto 3.0 % (w/w) evidencing minimization of catabolite repression in SSF. Babu and Satyanarayana (1995) reported only 5% reduction in amylase production by *B. coagulans* B49 even at addition of glucose concentrations as high as 20%, proving minimal repression effect.

8.4.3. Effect of Inoculum Size

Inoculum level plays a dominant role in enzyme synthesis; reduced level of inoculum may not be sufficient for startup of growth and enzyme synthesis on different substrates. However, increase in the number of cells ensures a rapid multiplication of biomass and enzyme synthesis. After a certain deadline, enzyme production could decrease because of depletion of nutrients due to the increased biomass, consequently resulting in a decrease in metabolic activity (Kashyap et al. 2002). A balance between the multiplying microbial biomass and available substrate material would enhance the enzyme yield (Pandey et al. 2000). Studies carried by Jha et al. (1995) indicated that maximum cellulase production was observed at spore concentrations of 1.0×10^4/gds and enzyme production was not increased even after further addition of inoculum strength. Kunamneni et al. (2005) achieved maximum productivity of amylase (534 U/g) by utilizing wheat bran as the solid substrate at an inoculum level of 10% (v/w). In serine protease production by *A. flavus* MTCC 9952 under SSF of wheat bran and corn cob mixture (1:1) an inoculum concentration of 1×10^8 spores/g substrate was found to be optimum (1621 U/g), however, further addition of inoculum retarded enzyme activity (Yadav et al. 2011).

Distinct patterns of chitinase yield with consideration to inoculum size were observed in isolates of *P. chrysogenum* (PPCS1 & PPCS2). Chitinase production increased linearly in case of PPCS2 with increase in inoculum size, however, in case of PPCS1 maximum chitinase yield was obtained at a spore concentration of 1×10^6/mL and further increase in inoculum size precisely decreased enzyme yield (Patidar et al. 2005). Suresh and chandrasekaran (1999) illustrated the usage of larger inoculum size critically affecting the yield of chitinase in *B. bassiana*. A linear increase in protease production was observed while using 10 to 100% (v/w) *B. subtilis* inoculum concentration. However, increase in the inoculum concentration from 100% to 300% resulted in a steady detoriation of enzyme production (Mukherjee et al. 2008).

8.4.4. Effect of pH

Hydrogen ion concentration of medium or surrounding environments is also one of the most important factors which contribute high production of enzymes. Prior et al. (1992) reported, using wheat straw as a substrate the optimum pH for maximum production of endoglucanase (1572 U/g) was at 4.0. Fungal cultures have been reported to give optimum enzyme production at pH 5.0 using various substrates (Ellaiah et al. 2002). At pH 5.0 *Aspergillus sp.* HA-2 showed maximum enzyme (264 U/gds) production in wheat bran with 10^6 spores/mL inoculum (Anto et al. 2006). Production of cellulase using mixed substrate (wheat bran and wheat straw, 1:9) did not vary appreciably within an initial pH range of 4.5–5.5 in *A. niger*. Interestingly, it was briefed that, in the course of time, maximum activity was obtained when the initial pH was adjusted to 4.5–5.0 (Jecu, 2000). The optimum pH for lignolytic enzymes, LiP and MnP production in SSF of *A. zapota* leaves by white rot fungi, *P. chrysosporium* was found to be at pH 5.0 (Ganesh Kumar et al. 2006). The optimum pH for cellulase production in the thermophilic fungus *T. aurantiacus* was found to be at pH 4.0 resulting in high endoglucanase (1572 U/g wheat straw) and β-glucosidase (101.6 U/g wheat straw) activities. Above pH 5.0, the growth of the fungus was very limited and as a consequence low enzyme yields were observed. These observations are in agreement with those reported in the literature confirming that filamentous fungi prefer acidic environments with optima at pH 3.8 - 5.0 (Ganesh Kumar et al. 2006).

8.4.5. Effect of Temperature

Temperature was the most important physical variable in SSF. This is particularly important parameter in SSF because it affects directly bioprocesses characteristics, spores germination, cell growth and the efficacy of the fermentation process. It has to be noted that the most used substrates for SSF affects the heat transfer making temperature as a critical parameter to regulate. Respiration during growth and metabolism depends on the oxygen consumption and CO_2 formation rate, which is highly exothermic and heat generation in turn relies on the level of biochemical activities of the microorganisms. The microbial growth is particularly sensitive to the rise or change of temperature and to the thermal gradients produced by heat generation. This makes temperature as one of the most critical issues controlling SSF. Maximum endoglucanase and β-glucosidase activities were obtained when the thermophilic fungus *T. aurantiacus* was grown at 50ºC (Grajek, 1986). The optimal temperature for the cellulase production of *A. terreus* M11 was 35–45ºC and the maximal cellulase production occurred at 45ºC. When the temperature was above 45ºC, the level of enzyme production gradually declined (Gao et al. 2008). Generally, the temperature maintained in the SSF system is in the range of 25-32ºC and it mainly depends on the growth kinetics of the microorganisms (Lonsane et al. 1985) rather than on the kind of enzyme produced. Krishna (1999) recorded high amount of cellulases production at 35ºC by *B. subtilis* (CBTK106), which is in line with bacterial temperature requirement.

8.4.6. Effect of Moisture Content

The moisture content of the medium has a critical significance to enzyme production in SSF. Moisture is either absorbed on the surface or complexed within the supporting solid matrix in SSF systems. High /low moisture content can cause imbalance in the transport phenomena at macroscopical and cellular level, but good diffusion of nutrients and gas exchange are partly granted by free water. Thus water content of the substrate plays an important role in both cell growth and enzyme production in SSF. Topakas et al. (2003) identified wheat straw with 80% moisture content as ideal substrate for feruloyl esterase production in thermophilic fungi *S. thermophile*. However, some researches proved optimal enzyme production at lower than 65%. The low enzyme activity at high substrate moisture levels could be accounted to the decreased porosity, alteration in particle structure, lower oxygen transfer or increased formation of aerial hyphae. Likewise, lower moisture levels lead to reduced diffusion of the nutrients in the substrate, decrease degree of swelling and higher water tension. The optimum moisture content of the solid medium for the production of phytase by *A. ficuum* NRRL 3135 is 64% (Ebune et al. 1995). Xylanase and pectinase production by *A. awamori* on grape pomace in SSF were high at 65% (w/w) initial moisture content. Both enzyme activities were less when the moisture content was higher or lower than 65%. Interestingly, Mukherjee et al. (2008) elucidated substrate (grass or potato peel) moisture content influence on protease production by *B. subtilis* DM-04 strain. With an increase in the initial moisture content of the substrate from 25% to 100% alkaline protease production was concomitantly enhanced and further increase in the moisture content of substrate resulted in a steady decline in protease yield. An increase in the initial moisture content of the mixed substrate (wheat bran and wheat straw, 1:9) from 55 to 74% greatly enhanced the endoglucanase production in *A. niger*. However, a further increase to 80% had a negative effect on the enzyme production (Jecu et al. 2000). Production of cellulolytic enzymes from the thermo acidophilic fungal *A. terreus* M11 under SSF of corn stover showed that increase of the initial moisture content from 60% to 80% greatly enhanced the enzyme activity in the broth, and till 80% the activity was the highest (Gao et al. 2008). This could be attributed to the faster growth of microorganism at high moisture content and the subsequent early initiation of the enzyme production. The higher moisture level enhanced fungal growth and cellulase production when lignocellulosic substrates were used as carbon sources in the SSF (Panagiotou et al. 2003). The highest inulinase activity in *K. marxianus* was obtained with SSF of sugarcane bagasse with >60% moisture content at 24h of fermentation (Astolfia et al. 2011). The lignolytic enzymes, LiP and MnP were produced by utilizing leaves of *A. zapota* with 60% moisture content by *P. chrysosporium* in SSF (Ganesh Kumar et al. 2006). The initial moisture content has been shown to affect the production of hydrolytic enzymes under SSF conditions by influencing the growth of the organism. Patidar et al. (2005) revealed wide difference with respect to moisture requirement for chitinase production in *P. chrysogenum* isolates (PPCS1 and PPCS2), where both exhibited maximum yield at 80 and 120% initial moisture, respectively. High level of water in the solid material could make the medium clumped, reduction in substrate porosity, change in particle structure, reduction of gas transfer and minimal diffusion which could result in decrease of enzyme production.

8.4.7. Influence of Supplementary Carbon and Nitrogen Sources

Selecting the best carbon and nitrogen sources and optimizing their concentrations increases the rates of enzyme production (Gomes et al. 2000). Kunamneni et al. (2005) showed enhanced amylase production (388 U/g) in SSF of soluble starch by the thermophilic fungus *T. lanuginosus*. Peptone increased amylase production (414 U/g) followed by tryptone and meat extract in SSF by *T. lanuginosus* (Kunamneni et al. 2005). Previous findings have shown that peptone, sodium nitrate and casein hydrolysate are good nitrogen supplements for amylase production in *A. niger* (Pandey et al. 1994) and *A. fumigatus* (Goto et al. 1998). Supplementation of sesamum oil cake with maltose was stimulatory for extra-cellular L-glutaminase synthesis by the yeast strain, *Z. rouxii* (10.92 in comparison to 8.53 U/gds) in SSF. However, different carbon sources such as glucose, sucrose, lactose, starch, sorbitol and mannitol supplementation on enzyme production retarded the yeast's activity of enzyme synthesis. Interestingly, supplementation of carbon sources to the wheat bran enhanced enzyme production by the yeast culture which was found to be maximum with glucose (8.14 U/gds compared to 3.99 U/gds in control) (Kashyap et al. 2002). Much similar result was observed by Sabu et al. (2005), when D-glucose at 0.5% w/v almost doubled the glutaminase yield by the marine fungus, *B. bassiana* under SSF on an inert substrate, polystyrene. Different nitrogen sources (ammonium sulphate, potassium nitrate, ammonium hydrogen phosphate, yeast extract, urea, peptone and corn steep liquor) showed different levels of cellulolytic enzymes production in *T. aurantiacus* under SSF of agricultural wastes, the results showed that enzyme activities are higher with inorganic nitrogen, ammonium sulphate (Kalogeris et al. 2003). These data are in contrast to those reported in the literature, where high enzymatic activities were observed when *T. aurantiacus* was grown on organic nitrogen sources (Gomes et al. 1994). Similarly, cellulolytic activities were higher with organic nitrogen, yeast extract, under SSF of corn stover using thermo acidophilic fungal *A. terreus* (Gao et al. 2008). The soybean meal medium supplemented with urea and soybean oil significantly increased the lipase production in *Penicillium* P58 and P74, 139.2 U/gds and 140.7 U/gds in 48h of fermentation (Rigo et al. 2010). Yadav et al. (2011) found in addition of organic nitrogen sources (50 mg/gss), only malt extract had an inductive effect on protease production (2923 U/gss) and in inorganic nitrogen sources (20 mg/gss), di-ammonium hydrogen orthophosphate reached maximum of 3640 U/gss in SSF of wheat bran and corn cob mixture (1:1), using *A. flavus* MTCC 9952. In an attempt to evaluate the effect of different nitrogen sources, Gutierrez-Correa et al. (1999) achieved Carboxymethylcellulase

(CMCase) activities of 65.8 and 51.4 U/gds after 48 h, respectively, by supplementing inorganic nitrogen source and soymeal to sugarcane bagasse in *A. niger* ATCC 10864.

8.5. Strain Improvement of Enzymes on SSF

During the past three to four decades, lot of efforts have been made for preparing mutagenesis and genetic modifications to retrieve reformed microbial strains capable of producing high levels of enzymes. One of the major problems with the mutagenesis (eg. usage of 3-nitro,5-methylguanidine and UV irradiation) is random mutation rather than the desired genetic change leading to either upregulation of desired trait or suppression of the essential functions of the cells, thereby lowering its production trait. In certain cases, the yields of product concentration were higher in SSF than in SmF. But, studies have shown that strain which showed good yield in SmF is failed to reproduce the same yield in SSF. This may be due to major complexity in SSF because of heterogeneity of medium and complex substrate. In order to enhance the enzyme production, improving the strains specific trait is required through the process of strain improvement. The process of strain improvement is much tedious which allows modifying the strain to achieve the desired trait or existing trait in an over expressed condition.

8.5.1. Mutation

Vu et al. (2011) developed a mutant strain of *A. niger* E-30 by bimutation breeding to enhance β-mannanase production by SSF. To obtain mutant strain, wild *A. niger* LW-1 was mutated by the compound mutagenesis of vacuum microwave (VMW) and ethyl methane sulfonate (EMS). The β-mannanase activity of the screened mutant *A. niger* E-30, cultivated at 32°C for 96 h, reached 36675 U/g dried koji and this activity was 1.98-fold higher than that (18501 U/g dried koji) of the parent strain. To produce cellulolytic enzyme efficiently, Liu et al. (2011) used *P. decumbens* strain L-06 to prepare mutant ML-017 using EMS and UV-irradiation. The maximum cellulase production (5.76 U/g) under the optimized condition was obtained in mutant strain ML-017 by SSF with rice bran as the solid substrate, which is 44.12% higher compared to its original strain. Vu et al. (2010) improved wild *Aspergillus* strain for the hyper-production of raw-starch-digesting enzyme by sequential exposures to gamma-irradiation of Co^{60}, UV, and treatments with N-methyl- N'-nitrosoguanidine. The improved mutant strain *Aspergillus* sp. XN15 produced starch-digesting enzyme 19.4 times greater that of the wild strains. Guo et al. 2009 isolated inulinase overproducers in marine yeast *P. guilliermondii* by mutation, using UV light and $LiCl_2$. In SSF culture of the mutant M-30, under the optimized conditions, enhanced inulinase production was noticed (455.9 U/gds). However, its parent strain produced only 291.0 U/gds of inulinase activity. Montiel-Gonzalez et al. (2004) obtained haploid mutant strains of *Aspergillus niger* Aw96-3 and Aw96-4 producing increased invertase than the parental strain *A. niger* C28B25. Diploid strains (AD96-4 and DAR2) grown on inert support using PUF under optimized conditions showed 5-18 times increased invertase productivity levels (2677 U and 1324 U) when compared to the haploid mutants Aw96-3 and Aw96-4 (254 U and 62 U) and parent strain

(441 U). Dwivedi et al. (2011) developed *P. oxalicum* SAU_E-3.510 through mixed mutagenesis of wild type *P. oxalicum* SA-8 ITCC 6024. Under optimized experimental conditions, co-cultivation of mutant *P. oxalicum* SAU_E-3.510 and *P. ostreatus* MTCC 1804 resulted in 58% and 33% higher levels of xylanase and laccase production, respectively. Mixed substrate combination of sugarcane bagasse and black gram husk in a ratio of 3:1 was found to be the most ideal solid substrate and support for fungal colonization and enzyme production during cocultivation. Moreover, maximum levels of xylanase (8205.31 U/g) and laccase (375.53 U/g) during SSF were obtained by using 4 g of solid support with 80% of moisture content. Important advantage of mutation is its effectiveness, since it rapidly increase the production trait and a major drawback is that it is labor intensive. Recently, these random screening methods have been replaced by molecular genetic strategies.

8.5.2. Protoplasts Fusion

Some researchers have elucidated genetic improvement of microorganisms for hyper enzyme production during solid state growth. Current focus of strain improvement strategies is towards protoplast fusion which finds broad spectra applications in pure and applied genetics. This technology is applied for developing interspecific, intraspecific and intrageneric suprahybrids with higher production trait than their parental strains. Protoplast fusion is a feasible method for inter-specific and inter-generic hybridization for strain improvement among fungus. Kavitha and Umesh-Kumar (2000) observed low production of pectinase in *A. carbonarius* growing on wheat bran solid substrate due to its fluctuating growth rates of the fungus. However, strain improvement strategy by protoplasts fusion influenced *A. niger* and higher growth rates were observed. Moreover, strain improvement technique for *A. carbonarius* and *A. niger* involving interspecific protoplast fusion was used to develop new *mutants* for obtaining higher pectinase production in solid state. One of the mutants strain screened and selected for higher activities of alpha-glucosidase showed growth rates improvement on wheat bran solid substrate coupled with increased pectinase production (Kavitha and Umesh-Kumar, 2000). El-Bondkly and El-Gendy (2012) isolated cellulolytic endophytic strains including *Trichoderma sp.* Merv6, *Penicillium sp.* Merv2 and *Aspergillus* sp. Merv70 from marine sponge *L. corticata*. A recombinant strain (Tahrir-25) was obtained from cellulolytic endophytic strains by mutagenesis and intergeneric protoplast fusion. This strain overproduced cellulases (exo-β-1,4-glucanase, endo-β-1,4-glucanase and β-1,4-glucosidase) that promoted complete cellulolysis of agricultural residues.

8.5.3. Alternative strain Improvement Strategies

An effective strain improvement procedure could aid in an economical enzyme production and thus increases its industrial utilization. Gene disruption studies on *Aspergillus oryzae* revealed significant increase in amylase and protease acitivity. Te Biesebeke et al. (2005) subjected the pclA gene encoding a secretion pathway specific endo-protease and pg/pi-tp gene encoding phosphatidylglycerol / phosphatidylinositol transfer protein to gene disruption to produce mutants of *A. oryzae*. The identified pclA and pg/pi-tp disrupted mutant strains when grown on SSF using wheat kernel under optimized condition produced at least 50%, 100% and 90% more productivity levels of amylase, glucoamylase and protease respectively, than wild type strain (Te Biesebeke et al. 2005). Besides the general strain improvement process by mutation and selection, other methods like protoplast fusion, amplification of secondary metabolite production genes, and inactivation of competing pathways are few of the genetic engineering procedures to improve the production trait in microorganisms (Barrios-Gonzalez et al. 2003).

8.6. Future Trends

Agricultural residues cause serious disposal problems especially in countries where the economy relies much on agriculture and its related activities. These wastes have trapped carbon reserves in plenty and they are easily assimilated by the microorganisms due to their organic nature. This makes such wastes very appropriate for their exploitation as raw materials in the production of enzymes under SSF conditions. One of the great potential of SSF was to utilize the solid wastes of agricultural, industrial and organics waste to convert into valuable metabolite products. These address two aspects, conversion of underutilized resources into value added products and appropriate utilization of solid wastes. In recent days, SSF is an economically viable and it finds application for the specific type of secondary metabolite production with free of contaminations which of interest to many area of advanced biology. The commercial scale-up process is much difficult, attributing to limitations in control of parameters such as temperature, nutrient concentration, pH, and moisture influencing fermentations, which are requisite to the system. To develop an economically stable SSF that can be scaled up to industrial level, production process must address the following stages: selection of appropriate microorganisms and substrates, fermentation optimization, upstream and downstream processing. In selecting a microorganism, the concept of using monoculture is shifting towards application of co-culture (wild and recombinant culture) due to better substrate utilization, increased productivity, and improved adaptability to changing conditions and has a superior resistance to contamination by unwanted microbes. Enzymes from microorganisms of higher yields with optimized condition are given much importance in increasing the production of the enzymes. Modern biotechnological techniques like *in vitro* mutagenesis, protoplasmic fusion, cloning and over expression of enzyme producing genes, fusion protein technologies gives hand to improve the yield of enzyme production to several folds and quality of the enzymes activity as well overcoming the usual fermentation inhibition like feedback and catabolite repression. Moreover, in recent years, there has been considerable interest in microbes and their products, which exist in extremophilic environments. Marine microorganisms or their bioactive enzymes may play important roles in processes that require high-salt concentrations. Microbial strains, which can grow under high salt condition, could be of special interest for the processes having high commercial value.

Marine microorganisms are highly adapted to survive in the harsh environment such as high pressure, low temperature, aphotic zone, limited organic food reserves, etc. The evolutionary causes and ecological consequences of genetic variation induced the marine microorganisms to generate multifarious biocatalytic systems to adapt and survive to the complicated extremophilic habitats. Therefore, marine microbial enzymes can offer unique enzymes with wide range of bioactive properties. Exploring marine microorganisms as a source for novel enzyme, utilization of appropriate support and use of sea water as fermentation media in SSF will gain broad industrial and pharmaceutical significance in near future.

ACKNOWLEDGMENTS

The authors gratefully acknowledge the financial support given by the Earth System Science Organization, Ministry of Earth Sciences, Government of India. The authors are thankful to Dr. M.A. Atmanand, Director, National Institute of Ocean Technology (NIOT), Ministry of Earth Sciences, Govt. of India, for his constant support and encouragement for preparation of this article.

REFERENCES

Agarwal D, Patidar P, Banerjee T, Shridhar P (2003). Production of alkaline protease by *Penicillium* sp. under SSF conditions and its application to soy protein hydrolysis. *Process Biochem* 1-6.

Aikat K, Bhattacharyya BC (2000). Protease extraction in solid state fermentation of wheat bran by a local strain of *Rhizopus oryzae* and growth studies by the soft gel technique. *Process Biochem* 35:907-914.

Anisha GS, Sukumaran RK, Prema P (2008). Evaluation of alpha-galactosidase biosynthesis by *Streptomyces griseoloalbus* in solid-state fermentation using response surface methodology. *Lett Appl Microbiol* 46:338-343.

Anto H, Trivedi UB, Patel KC (2006). Glucoamylase production by solid-state fermentation using rice flake manufacturing waste products as substrate. *Bioresour Technol* 97:1161-1166 .

Arantes V, Silva EM, Milagres AMF (2011). Optimal recovery process conditions for manganese-peroxidase obtained by solid-state fermentation of eucalyptus residue using *Lentinula edodes*. *Biomass Bioenergy* 35:4040-4044.

Asgher M, Iqbal HM, Irshad M (2012). Characterization of purified and xerogel immobilized novel lignin peroxidase produced from *Trametes versicolor* IBL-04 using solid state medium of corncobs. *BMC Biotechnol* 12:46.

Asha Poorna C, Prema P (2006). Production and partial characterization of endoxylanase by *Bacillus pumilus* using agro industrial residues. *Biochem Eng J* 32:106-112.

Ashok Kumar B, Kayalvizhi N, Gunasekaran P (2001). Optimization of media for ß-fructofuranosidase production by *Aspergillus niger* in submerged and solid state fermentation, *Process Biochem* 36:1241-1247.

Asther M, Haon M, Roussos S, Record E, Delattre M, Lesage-Meessen L (2002). Feruloyl esterase from *Aspergillus niger*: A comparison of the production in solid state and submerged fermentation, *Process Biochem* 38:685-691.

Astolfia V, Jorisa J, Verlindoa R, Oliveiraa J V, Maugerib F, Mazuttic MA, de Oliveiraa D, Treichel H (2011). Operation of a fixed-bed bioreactor in batch and fed-batch modes for production of inulinase by solid-state fermentation. *Biochem Eng J* 58-59:39-49.

Auria R, Hernandez S, Raimbault M, Revah S (1990). Ion exchange resin: a model support for solid state growth fermentation of *Aspergillus niger*. *Biotechnol Tech* 4:391-396.

Babu KR, Satyanarayana T (1995). α-Amylase production by thermophilic *Bacillus coagulans* in solid state fermentation. *Process Biochem* 30:305-309.

Bano S, Qader SA, Aman A, Syed MN, Durrani K (2013). High production of cellulose degrading endo-1,4-β-d-glucanase using bagasse as a substrate from *Bacillus subtilis* KIBGE HAS. *Carbohydr Polym* 91:300-304.

Barrios-Gonalez J, Fernandez FJ, Tomasini A (2003). Microbial secondary metabolites production and strain improvement. *Ind J Biotechnol* 2:322-333.

Baysal Z, Uyar F, Aytekin C (2003). Solid state fermentation for production of amylase by a thermotolerant *Bacillus subtilis* from hot spring water. *Process Biochem* 38:1665-1668.

Bennett P, Yang ST (2012). Beneficial effect of protracted sterilization of lentils on phytase production by *Aspergillus ficuum* in solid state fermentation. *Biotechnol Prog* 28:1263-1270.

Chaari F, Kamoun A, Bhiri F, Blibech M, Ellouze-Ghorbel R, Ellouz-Chaabouni S (2012). Statistical optimization for the production of lichenase by a newly isolated *Bacillus licheniformis* UEB CF in solid state fermentation using pea pomace as a novel solid support. *Ind Crop Prod* 40:192-198.

Chatterjee R, Dutta A, Banerjee R, Bhattacharyya BC (1996). Production of tannase by solid-state fermentation. *Bioprocess Eng* 14:159-162.

Chutmanop J, Chuichulcherm S, Chisti Y, Srinophakun P (2008). Protease production by *Aspergillus oryzae* in solid-state fermentation using agroindustrial substrates. *Chem Technol Biotechnol* 83:1012–1018.

Dahiya N, Tewari R, Tiwari RP, Hoondal GS (2005). Production of an antifungal chitinase from *Enterobacter* sp. NRG4 and its application in protoplast production. World J Microbiol Biotechnol 21:1611-1616.

De Azeredo LAI, Freire DMG, Soares RMA, Leite SGF, Coelho RRR (2004). Production and partial characterization of thermophilic proteases from *Streptomyces sp.* isolated from Brazilian cerrado soil. *Enzyme Microb Technol* 34:354-358.

Dwivedi P, Vivekanand V, Pareek N, Sharma A, Singh RP (2011). Co-cultivation of mutant *Penicillium oxalicum* SAUE-3.510 and *Pleurotus ostreatus* for simultaneous biosynthesis of xylanase and laccase under solid-state fermentation. *N Biotechnol* 28:616-626.

Ebune A, AI-Asheh S and Duvnjak Z (1995). Production of phytase during solid state fermentation using *Aspergillus ficuum* nrrl 3135 in canola meal. *Bioresour Technol* 53:7-12.

El-Bondkly AM, El-Gendy MM (2012). Cellulase production from agricultural residues by recombinant fusant strain of a fungal endophyte of the marine sponge *Latrunculia corticata* for production of ethanol. *Antonie van Leeuwenhoek* 101:331-346.

Elibol M, Moreira AR (2005). Optimizing some factors affecting alkaline protease production by a marine bacterium *Teredinobacter turnirae* under solid substrate fermentation. *Process Biochem* 40:1951-1956.

Ellaiah P, Adinarayana K, Bhavani Y, Padmaja P, Srinivasula B (2002). Optimization of process parameters for glucoamylase production under solid state fermentation by a newly isolated *Aspergilus* species. *Process Biochem* 38:615-620.

Fernandez-Lahore HM, Fraile ER, Cascone O (1998). Acid protease recovery from a solid-state fermentation system. *J Biotechnol* 62:83-93.

Ferreira V, da Silva R, Silva D, Gomes E (2010). Production of pectate lyase by *Penicillium viridicatum* RFC3 in solid-state and submerged fermentation. *Int J Microbiol* 1-8 .

Ganesh Kumar A, Sekaran G, Krishnamoorthy S (2006). Solid state fermentation of *Achras zapota* lignocellulose by *Phanerochaete chrysosporium*. *Bioresour Technol* 97:1521-1528.

Gao J, Weng H, Zhu D, Yuan M, Guan F, Xi Y (2008). Production and characterization of cellulolytic enzymes from the thermoacidophilic fungal *Aspergillus terreus* M11 under solid-state cultivation of corn stover. *Bioresour Technol* 99:7623-7629.

Gautam P, Sabu A, Pandey A, Szakacs G, Soccol CR (2002). Microbial production of extra-cellular phytase using polystyrene as inert solid support. *Bioresour Technol* 83:229-233.

Gessesse A, Mamo G (1999). High-level xylanase production by an alkaliphilic *Bacillus sp.* by using solid-state fermentation. *Enzyme Microb Technol* 25:68-72.

Gomes DJ, Gomes J, Steiner W (1994). Factors influencing the induction of endo-xylanase by *Thermoascus aurantiacus*. *J Biotechnol* 33:87-94.

Gomes I, Gomes J, Gomes DJ, Steiner W (2000). Simultaneous production of high activities of thermostable endoglucanase and betaglucosidase by the wild thermophilic fungus *Thermoascus aurantiacus*. *Appl Microbiol Biotechnol* 53:461-468.

Goto CE, Barbosa EP, Kistner LCL, Gandra RF, Arrias VL, Peralta RM (1998). Production of amylase by *Aspergillus fumigatus*. *Rev Microbiol* 29:99-103.

Grajek W (1986). Temperature and pH optima of enzyme activities produced by cellulolytic fungi in batch and solid state cultures. *Biotechnol Lett* 8:587-590.

Graminha EBN, Gonc alves, AZL, Pirota RDPB, Balsalobre MAA, da Silva R, Gomes E (2008). Enzyme production by solid-state fermentation: Application to animal nutrition. *Anim Feed Sci Technol* 144:1-22.

Guo N, Gong F, Chi Z, Sheng J, Li J (2009). Enhanced inulinase production in solid state fermentation by a mutant of the marine yeast *Pichia guilliermondii* using surface response methodology and inulin hydrolysis. *J Ind Microbiol Biotechnol* 36:499-507.

Gupta U, Kar R (2009). Xylanase Production by a thermo-tolerant *Bacillus* species under solid-state and submerged fermentation. *Braz Arch Biol Technol* 52:1363-1371.

Gutierrez-Correa M., Portal L, Moreno P, Tengerdy RP (1999). Mixed culture solid substrate fermentation of *Trichoderma reesei* with *Aspergillus niger* on sugarcane bagasse. *Bioresour Technol* 68:173-178.

Heerd D, Yegin S, Tari C, Fernandez-Lahore M (2012). Pectinase enzyme-complex production by *Aspergillus* spp. in solid-state fermentation: A comparative study. *Food Bioprod Process* 90:102-110.

Hesseltine CW (1987). Solid state fermentation – An overview. *Int Biodeterior* 23:79-89.

Holker U, Hofer M, Lenz J (2004). Biotechnological advantages of laboratory-scale solid state fermentation with fungi. *Appl Microbiol Biotechnol* 64:175-186.

Hu T, Zhoub Y, Daic L, Wanga Y, Liuc D, Zhangb J, Liub H (2011). Enhanced cellulase production by solid state fermentation with polyurethane foam as inert supports. *Procedia Eng* 18:335-340.

Iyer P, Singhal RS (2010). Production of glutaminase (E.C. 3.2.1.5) from *Zygosaccharomyces rouxii* in solid-state fermentation and modeling the growth of *Z. rouxii* therein. *J Microbiol Biotechnol* 20:737-748.

Jecu L (2000). Solid state fermentation of agricultural wastes for endoglucanase production. *Ind Crop Prod* 11:1-5.

Jha K, Khare SK, Gandhi AP (1995). Solid-state fermentation of soyhull for the production of cellulase. *Bioresour Technol* 54:321-322.

Kalogeris E, Christakopoulos P, Katapodis P, Alexiou A, Vlachou S, Kekos D, Macris BJ (2003). Production and characterization of cellulolytic enzymes from the thermophilic fungus *Thermoascus aurantiacus* under solid state cultivation of agricultural wastes. *Process Biochem* 38:1099-1104.

Kammoun R, Farhat A, Chouayekh H, Bouchaala K, Bejar S (2012). Phytase production by *Bacillus subtilis* US417 in submerged and solid state fermentations. *Annals Microbiol* 62:155-164.

Kamra P, Satyanarayana T (2004). Xylanase production by the thermophilic mold *Humicola lanuginosa* in solid-state fermentation. *Appl Biochem Biotechnol.* 119:145-157 .

Karp SG, Faraco V, Amore A, Birolo L, Giangrande C, Soccol VT, Pandey A, Soccol CR (2012). Characterization of laccase isoforms produced by *Pleurotus ostreatus* in solid state fermentation of sugarcane bagasse. *Bioresour Technol* 114:735-739.

Kashyap P, Sabu A, A Pandey, Szakacs G, Soccol CR (2002). Extra-cellular L-glutaminase production by *Zygosaccharomyces rouxii* under solid-state fermentation. *Process Biochem* 38:307-312.

Kavitha R, Umesh-Kumar S (2000). Genetic improvement of *Aspergillus carbonarius* for pectinase overproduction during solid state growth. *Biotechnol Bioeng* 67:121-125.

Khandeparker R, Numan MTH, Mukherjee B, Satwekar A, Bhosle NB (2008). Purification and characterization of α-L-arabinofuranosidase from *Arthrobacter sp.* MTCC 5214 in solid state fermentation. *Process Biochem* 43:207–212.

Kirk O, Borchert TV, Fuglsang CC (2002). Industrial enzyme applications. *Curr Opin Biotechnol* 13(4):345-351.

Krishna C (1999). Production of bacterial cellulases by solid state bioprocessing of banana wastes. *Bioresour Technol* 69:231-239.

Krishna C, Chandrasekaran M (1996). Banana waste as substrate for α-amylase production by *Bacillus subtilis* (CBTK 106) under solid-state fermentation. *Appl Microbiol Biotechnol* 46:106-111.

Kumar R, Sharma J, Singh R (2007). Production of tannase from *Aspergillus ruber* under solid-state fermentation using jamun (*Syzygium cumini*) leaves. *Microbiol Res* 162: 384-390.

Kunamneni A, Permaul K, Singh S (2005). Amylase Production in solid state fermentation by the thermophilic fungus *Thermomyces lanuginosus*. *J Biosci Bioeng* 100:168-171.

Lakshmi GS, Rao CS, Rao RS, Hobbs PJ, Prakasham RS (2009). Enhanced production of xylanase by a newly isolated *Aspergillus terreus* under solid state fermentation using palm industrial waste: A statistical optimization. *Biochem Eng J* 48:51-57.

Liu YT, Luo ZY, Long CN (2011). Cellulase production in a new mutant strain of *Penicillium decumbens* ML-017 by solid state fermentation with rice bran. *N Biotechnol* 28:733-737.

Lonsane BK, Ghildyal NP, Budiatman S, Ramakrishna SV (1985). Engineering aspects of solid state fermentation. *Enzyme Microb Technol* 7:258–265.

Madeira JV, Macedo JA, Macedo GA (2012). A new process for simultaneous production of tannase and phytase by *Paecilomyces variotii* in solid-state fermentation of orange pomace. *Bioprocess biosys eng* 35:477-482.

Madhuri A, Nagaraju B, Harikrishna N, Reddy G (2012). Production of alkaline protease by *Bacillus altitudinis* GVC11 using castor husk in solid-state fermentation. *Appl Biochem Biotechnol* 167:1199-1207.

Mahadik ND, Puntambekar US, Bastawde KB, Khire JM, Gokhale DV (2002). Production of acidic lipase by *Aspergillus niger* in solid state Fermentation. *Process Biochem* 38:715-721.

Mahanta N, Gupta A, Khare SK (2008). Production of protease and lipase by solvent tolerant *Pseudomonas aeruginosa PseA* insolid-state fermentation using *Jatropha curcas* seed cake as substrate. *Bioresour Technol* 99:1729-1735.

Maldonado MC, Strasser de Saad AM (1998). Production of pectinesterase and polygalacturonase by *Aspergillus niger* in submerged and solid state systems. *J Ind Microbiol Biotechnol* 20:34–38.

Maller A, Damasio AR, da Silva TM, Jorge JA, Terenzi HF, Polizeli M de LT de Ms (2011). Biotechnological potential of agro-industrial wastes as a carbon source to thermostable polygalacturonase production in *Aspergillus niveus*. *Enzyme Res,* ID 289206, 1-6.

Marin-Cervantes MDC, Matsumoto Y, Ramirez-Coutino L, Rocha-Pino Z, Viniegra G, Shirai K (2008). Effect of moisture content in polyurethane foams as support for solid-substrate fermentation of *Lecanicillium lecanii* on the production profiles of chitinases. *Process Biochem* 43:24–32.

Martins ES, Silva D, Da Silva R, Gomes E (2002). Solid-state production of thermostable pectinases from thermophilic *Thermoascus aurantiacus*. *Process Biochem* 37:949–954.

Masui DC, Zimbardi AL, Souza FH, Guimaraes LH, Furriel RP, Jorge JA (2012). Production of a xylose-stimulated β-glucosidase and a cellulase-free thermostable xylanase by the thermophilic fungus *Humicola brevis* var. *thermoidea* under solid state fermentation. *World J Microbiol Biotechnol* 28:2689-2701.

Mazutti MA, Bender JP, Treichel H, Di Luccio M (2006) Optimization of inulinase production by solid state fermentation using sugarcane bagasse as substrate. *Enzyme Microb Technol* 39:56–59.

Miron J, Vazquez JA, Gonzalez P, Murado MA (2010). Enhancement glucose oxidase production by solid-state fermentation of *Aspergillus niger* on polyurethane foams using mussel processing wastewaters. *Enz Microb Technol* 46:21-27.

Mohan Kumar NS, Ramasamy R, Manonmani HK (2013). Production and optimization of l-asparaginase from *Cladosporium* sp. using agricultural residues in solid state fermentation. *Ind Crop Prod* 43:150– 158.

Montiel-Gonzalez AM, Viniegra-Gonzalez G, Fernandez FJ, Loera O (2004). Effect of water activity on invertase production by improved diploids strains of *Aspergillus niger*. *Process Biochem* 39:2085-2090.

Mukherjee AK, Adhikari H, Rai SK (2008). Production of alkaline protease by a thermophilic *Bacillus subtilis* under solid-state fermentation (SSF) condition using *Imperata cylindrical* grass and potato peel as low-cost medium: Characterization and application of enzyme in detergent formulation. *Biochem Eng J* 39:353-361.

Murado MA, Gonzfilez MP, Torrado A, Pastrana LM (1997). Amylase production by solid state culture of *Aspergillus oryzae* on polyurethane foams. Some mechanistic approaches from an empirical model. *Process Biochem* 32:35-42.

Nandakumar MP, Thakur MS, Raghavarao KSMS, Ghildyal NP (1999). Studies on catabolite repression in solid state fermentation for biosynthesis of fungal amylases. *Lett appl microbiol,* 29:380-384.

Narra M, Dixit G, Divecha J, Madamwar D, Shah AR (2012). Production of cellulases by solid state fermentation with *Aspergillus terreus* and enzymatic hydrolysis of mild alkali-treated rice straw. *Bioresour Technol* 121:355-361.

Nigam P, Singh D (1994). Solid-state (substrate) fermentation systems and their applications in biotechnology. *J Basic Microbiol* 34:405-423.

Panagiotou G, Kekos D, Macris BJ, Christakopoulos P (2003). Production of cellulolytic and xylanolytic enzymes by *Fusarium oxysporum* grown on corn stover in solid state fermentation. *Ind Crop Prod* 18:37-45.

Pandey A (2003). Solid-state fermentation. *Biochem Eng J* 13:81-84.

Pandey A, Ashakumary L, Selvakumar P, Vijyalakshmi KS (1994). Influence of water activity on growth and activity of *Aspergillus niger* for glycoamylase production in solid-state fermentation. *World J Microbiol Biotechnol,* 10: 485-486.

Pandey A, Soccol CR, Mitchell DA (2000). New developments in solid-state fermentation. I. Bioprocesses and products. *Process Biochem* 35:1158-1169.

Pandey A, Soccol CR, Rodriguez-Leon JA, Nigam P (2001). History and development of solid state fermentation. In: *Solid-state fermentation in biotechnology: fundamentals and applications Pandey* A. (Ed.) Asiatech Publishers, New Delhi 3-7.

Papinutti VL, Forchiassin F (2007). Lignocellulolytic enzymes from *Fomes sclerodermeus* growing in solid-state fermentation. *J Food Eng* 81:54–59.

Patidar P, Agrawal D, Banerjee T, Patil S (2005). Optimisation of process parameters for chitinase production by soil isolates of *Penicillium chrysogenum* under solid substrate fermentation. *Process Biochem* 40:2962–2967.

Patil SR, Dayanand A (2006). Optimization of process for the production of fungal pectinases from deseeded sunflower head in submerged and solid-state conditions. *Bioresour Technol* 97:2340–2344.

Prabhu GN, Chandrasekaran M (1995) Polystyrene: an inert carrier for L-glutaminase production by marine *Vibrio costicola* under solid-state fermentation. *World J Microbiol Biotechnol* 11:683–684.

Prabhu GN, Chandrasekaran M (1997). Impact of process parameters on l-glutaminase production by marine *Vibrio costicola* in solid state fermentation using polystyrene as an inert support. *Process Biochem* 32:285-289.

Prakasham RS, Subba Rao Ch, Sreenivas Rao R, Sarma PN (2005). Alkaline protease production by an isolated *Bacillus circulans* under solid-state fermentation using agroindustrial waste: process parameters optimization. *Biotechnol Prog* 21:1380−1388.

Prior BA, Du Preez JC, Rein PW (1992). Environmental parameters. In: Doelle H, Mitchell D, Rolz C, editors. In *solid substrate cultivation.* London and New York: Elsevier Appl Sci 1-16.

Puri S, Beg QK, Gupta R (2002). Optimization of alkaline protease production from *Bacillus sp.* by response surface methodology. *Curr Microbiol* 44:286–290.

Raj V, Hemashenpagam N (2012). Production and medium optimization of amylase by *Bacillus* using fermentation methods. *J Microbiol Biotech Res* 2:481-484.

Ramachandran S, Fontanille P, Pandey A, Larroche C (2008). Stability of glucose oxidase activity of *Aspergillus niger* spores produced by solid-state fermentation and their role as biocatalysts in bioconversion reaction. *Food Technol Biotechnol* 46:190–194.

Reddy GV, Ravindra Babu P, Komaraiah P, Roy KRRM, Kothari IL (2003). Utilization of banana waste for the production of lignolytic and cellulolytic enzymes by solid substrate fermentation using two *Pleurotus* species (*P. ostreatus* and *P. sajor-caju*). *Process Biochem* 38:1457-1462.

Renovato J, Gutierrez-Sanchez G, Rodriguez-Duran LV, Bergman C, Rodriguez R, Aguilar CN (2011). Differential properties of *Aspergillus niger* tannase produced under solid-state and submerged fermentations. *Appl Biochem Biotechnol* 165:382-395.

Rigo E, Ninow JL, Lucciob MD, Oliveira JV, Vardanega R, de Oliveira D, Treichel H (2010). Lipase production by solid fermentation of soy bean meal with different supplements. *LWT- Food Sci Technol* 43:1132-1137.

Rodriguez A, Mateos JC, Nungaray J, Gonzaez V, Bhagnagar T, Roussos S, Cordova J, Baratti J (2006). Improving lipase production by nutrient source modification using *Rhizopus homothallicus* cultured in solid state fermentation. *Process Biochem* 41: 2264-2269.

Rodriguez-Couto S (2012). A promising inert support for laccase production and decolouration of textile wastewater by the white-rot fungus *Trametes pubescesns. J Hazard Mater* 233–234:158-162.

Rodriguez-Couto S, Gundin M, Lorenzo M, Sanroman MA (2002). Screening of supports and inducers for laccase production by *Trametes versicolor* in semi-solid-state conditions. *Process Biochem* 38:249-255.

Sabu A, Keerthi TR, Rajeev Kumar S, Chandrasekaran M (2000). L-Glutaminase production by marine *Beauveria* sp. under solid state fermentation. *Process Biochem* 35:705-710.

Sabu A, Pandey A, Jaafar Daud M, Szakacs G (2005). Tamarind seed powder and palm kernel cake: two novel agro residues for the production of tannase under solid state fermentation by *Aspergillus niger* ATCC 16620. *Bioresour Technol* 96:1223-1228.

Sandhya C, Sumantha A, Szakacs G, Pandey A (2005). Comparative evaluation of neutral protease production by *Aspergillus oryzae* in submerged and solid-state fermentation. *Process Biochem* 40:2689-2694.

Sangeetha PT, Ramesh MN, Prapulla SG (2004). Production of fructosyl transferase by *Aspergillus oryzae* CFR 202 in solid-state fermentation using agricultural by-products. *Appl Microbiol Biotechnol* 65:530-537.

Shaheen I, Bhatti HN, Ashraf T (2008). Production, purification and thermal characterisation of invertase from a newly isolated *Fusarium sp.* under solid-state fermentation. *Int J Food Sci Technol* 43:1152-1158.

Silva D, Martins ES, Leite RSR, Da Silva R, Ferreira V, Gomes E (2007). Purification and characterization of an exo-polygalacturonase produced by *Penicillium viridicatum* RFC3 in solid-state fermentation. *Process Biochem* 42:1237-1243.

Singhvi MS, Adsul MG, Gokhale DV (2011). Comparative production of cellulases by mutants of *Penicillium janthinellum* NCIM 1171 and its application in hydrolysis of Avicel and cellulose. *Bioresour Technol* 102:6569-6572.

Souza DT, Bispo AS, Bon EP, Coelho RR, Nascimento RP (2012). Production of thermophilic endo-β-1,4-xylanases by *Aspergillus fumigatus* FBSPE-05 using agro-industrial by-products. *Appl Biochem Biotechnol* 166:1575-1585.

Srinivasan C, D'Souza TM, Boominathan K, Reddy CA (1995). Demonstration of laccase in the white rot basidiomycete *Phanerochaete chrysosporium* BKM-F-1767. *Appl Environ Microbiol* 61:4274-4277.

Sukumaran RK, Singhania RR, Pandey A (2005) Microbial cellulases – production, applications and challenges. *J Sci Ind Res* 64:832-844.

Sun SY, Xu Y (2008). Solid-state fermentation for 'whole-cell synthetic lipase' production from *Rhizopus chinensis* and identification of the functional enzyme. *Process Biochem* 43:219-224.

Sun SY, Xu Y (2009). Membrane-bound 'synthetic lipase' specifically cultured under solid-state fermentation and submerged fermentation by *Rhizopus chinensis*: A comparative investigation. *Bioresour Technol* 100:1336-1342.

Suresh PV, Chandrasekaran M (1999). Impact of process parameters on chitinase production by an alkalophilic marine *Beauveria bassiana* in solid state fermentation. *Process Biochem* 34:257-267.

Te Biesebeke R, Record E, van Biezen N, Heerikhuisen M, Franken A, Punt PJ, van den Hondel CA (2005). Branching mutants of *Aspergillus oryzae* with improved amylase and protease production on solid substrates. *Appl Microbiol Biotechnol* 69:44-50.

Topakas E, Kalogeris E, Kekos D, Macris BJ, Christakopoulos P (2003). Production and partial characterisation of feruloyl esterase by *Sporotrichum thermophile* in solid-state fermentation. *Process Biochem* 38:1539-1543.

Ustok FI, Tari C, Gogus N (2007). Solid-state production of polygalacturonase by *Aspergillus sojae* ATCC 20235. *J Biotechnol* 127:322-334.

Vu VH, Pham TA, Kim K (2010). Improvement of fungal strain by repeated and sequential mutagenesis and optimization of solid state fermentation for the hyper-production of raw-starch-digesting enzyme. *J Microbiol Biotechnol* 20:718-726.

Vu VH, Pham TA, Kim K (2011) Improvement of fungal cellulose production by mutation and optimization of solid state fermentation. *Mycobiology* 39:20–25.

Waghmare SR, Kulkarni SS, Ghosh JS (2011). Chitinase production in solid-state fermentation from *Oerskovia xanthineolytica* NCIM 2839 and its application in fungal protoplasts formation. *Curr Microbiol* 63:295-299.

Yadav SK, Bisht D, Shikha, Darmwal NS (2011) Oxidant and solvent stable alkaline protease from *Aspergillus flavus* and its characterization. Afr J Biotechnol 10:8630-8640.

Zhang C, Kim SK (2010). Research and application of marine microbial enzymes: status and prospects. *Mar Drugs* 8:1920-1934.

Zhu Y, Knol W, Smits PJ, Bol J (1996). Medium optimization for nuclease P1 production by *Penicillium citrinum* in solid-state fermentation using polyurethane foam as inert carrier. *Enz Microb Technol* 18:108-112.

In: Enzymes in Value-Addition of Wastes
Editors: Satinder Kaur Brar and Mausam Verma
ISBN: 978-1-62948-968-1

Chapter 4

MANGO AND LEGUME PROCESSING INDUSTRY BY-PRODUCTS: A POTENTIAL SOURCE OF VALUABLE ENZYMES

C. M. Ajila and U. J. S. Prasada Rao*
Department of Biochemistry and Nutrition
Central Food Technological Research Institute, Mysore, India

ABSTRACT

Food industry generates large quantities of waste by-products during harvesting and further processing of raw materials to consumable food products. Currently, increased efforts are made for better utilization of waste by-products, which results in reduction of environmental pollution and economic benefits. Food industry waste is plant/animal based product and is natural and safe source for nutrients and other valuable compounds such as colors, flavours, phenolics, enzymes etc. Agro- industry by-products can be converted into products with higher economic value than the main products through new scientific and technological methods. Therefore, isolation of valuable compounds from food industry waste products receiving increased attention.

Mango is one of the important tropical fruits in the world and India is the largest producer of mango. One of the major problems in mango industry is sap-injury which is characterized as browning or darkening of peel due to its contact with sap (latex). For controlling sap-injury, one of the oldest yet widely used methods is to harvest the mangoes with intact pedicel (stalk) and de-sap manually. Thus, sap is a waste by-product and found to be a good source of enzymes such as peroxidase (POD), polyphenol oxidase (PPO), lipoxygenase, protease and amylase. Peel is another by-product from mango processing industry and found to be rich source of oxidative and hydrolyzing enzymes like PPO, POD, protease, amylase and xylanase.

Legumes are widely grown in the world. Many legume seeds are subjected to primary processing before they are used for food preparation. During processing of legumes, about 25% is obtained as waste by-product. This by-product obtained was

* Corresponding author. Dr. U.J.S. Prasada Rao. Department of Biochemistry and Nutrition, CSIR-Central Food Technological Research Institute, Myosre-570 020, Inida. Tel.: +91 0821 2514876; fax: +91 0821 2517233. E-mail address: prasadarao_ummiti@yahoo.com.

separated into husk or hull, aleurone rich husk, plumule and germ fractions. Most of these fractions were found to be rich in oxido–reductases.

There has been an increasing trend toward the utilization of the solid-state fermentation (SSF) technique to produce several bulk chemicals and enzymes. Mango and legume processing by-products are found to be rich in nutrients and can be used as fermentation substrates for the production of enzymes. This chapter provides a brief discussion on the utilization of mango and legume processing wastes as a valuable source for the isolation of enzymes using conventional biochemical extraction methods and production of enzymes from microorganisms using fermentation technologies.

Thus, the waste by-products mentioned above are all rich in oxido-reductases and hydrolyzing enzymes which find potential applications in food processing, biomedical and health.

Keywords: Waste utilization, Mango, Legumes, Enzymes

INTRODUCTION

Food processing is essential to process surplus agricultural produce, to enhance the shelf-life of perishable items, and to increase the availability of seasonal fruit products throughout the year and also improve the nutritional quality of food by inactivating the anti-nutrients and toxins. However, one of the problems in food processing is that large quantities of waste by-products are generated, which causes disposal and environmental problems. Globally, 998 million tonnes of agricultural waste by-products are generated, which causes disposal and environmental problems (www.uncrd.or.jp).

Fruits and vegetables, viz., mango, citrus, pineapple, apple, grape, sugarcane and potato processing industries generate a large amount of waste. For example, during pine apple juice manufacturing, wastes consisting of stem, peel, core and crown are generated. In apple processing industry, peel, pomace, seed are the major by-products. At present, most of the agro by-products are disposed in the landfill, dumping in low lying areas or incineration, which pose serious environmental and health risks. For example, inappropriate management of landfill results in emissions of methane and carbon dioxide gases, and incineration releases dioxins, furans, acid gases as well as particulates [1,2].

Agricultural waste is a rich source of phytochemicals and enzymes [3-5]. Agro by-products can be used for the production of a number of valuable products, such as food, feed, fuel, and a variety of chemicals and bioactive compounds. Isolation of valuable compounds from inexpensive and readily available agro-food industry waste will provide useful phytochemicals, bioactive chemicals and enzymes at a lesser price as the raw materials are cheaper and these isolated compounds can be used in food, pharmaceutical, cosmetics and other industries [6].

Although the quantities of enzymes are very less compared to carbohydrates and biologically inactive proteins, they have high catalytic efficiency even at low concentrations and they are environmental friendly and catalyze highly specific reactions compared to chemical catalysts. Enzymes improve color, appearance, aroma and texture of foods. Some of the enzymes like peroxidase, laccase, protease and lipases are used for in food processing, pesticide degradation, clinical applications, digestive aid, production of bioactive peptides, biodiesel and water purification. Nowadays, enzymes are applied in the manufacturing of food, animal feed, detergents, pulp and paper, clothing and fine chemicals.

Thus, use of food industry waste not only provides certain raw materials at a cheaper cost but also minimizes the environmental pollution. Hence, agro-industrial waste can be considered as a new source for the production of value-added compounds. This chapter covers a brief overview of generated by-products of mango and legume and their potential application for isolation of the valuable compounds specifically enzymes, which can be used for further industrial applications.

Enzymes in Food Processing Waste

Agro-industrial wastes contain organic substances, such as polysaccharides, proteins, sugars, lipids, acids, starch, other nutrients, such as vitamins, minerals, fibers, and nutraceuticals like polyphenols and carotenoids. They are also good source of enzymes. Enzymes are important biological catalysts for many industrial and medical applications. Global enzyme demand will rise 6.3% annually through 2013 depending on the strong demand in the speciality enzyme segment and good growth in different industrial applications (www.freedoniagroup.com). North America and Western Europe have the healthy enzyme market and developing countries also showed a faster growth in the same sector. In recent years, efforts are being made to isolate enzymes from waste by-products using conventional biochemical methods and fermentation processes for the production of enzymes using microorganisms.

Pine apple crown contains highest activity of bromelain, a protease [7]. Mango peel and mango sap (latex) contain different types of enzymes [3,8-12]. Enzymes such as cellulases [13], amylolytic enzymes like amylase and glucoamylase [14,15], xylanases [16,17], polygalacturonase [18], fructosyl transferase [19], and lignolytic enzymes like laccase and manganese peroxidase [20-22] were produced by the solid state fermentation with different microorganisms, such as fungi, bacteria and yeast by food processing industrial waste.

Enzymes from Mango Processing Industry By-products

Mango (*Mangifera indica* L.) is an important tropical fruit and it ranks 5th among the fruit production in the world. Mango fruits are produced by several countries worldwide, however, Asia contributes approximately 77% of global mango production, and the Americas and Africa account for approximately 13% and 9%, respectively [23]. Among the mango producing countries, India ranks first contributing 40% of the world production. As mango is a seasonal fruit, it is processed into various products, such as pulp, pickle, amchur among others. During mango harvesting, sap (latex) is a by-product, whereas during mango processing, peel and kernel are waste by-products (Figure 1). Peel and sap contains several enzymes, which have various applications in food industry and other industries.

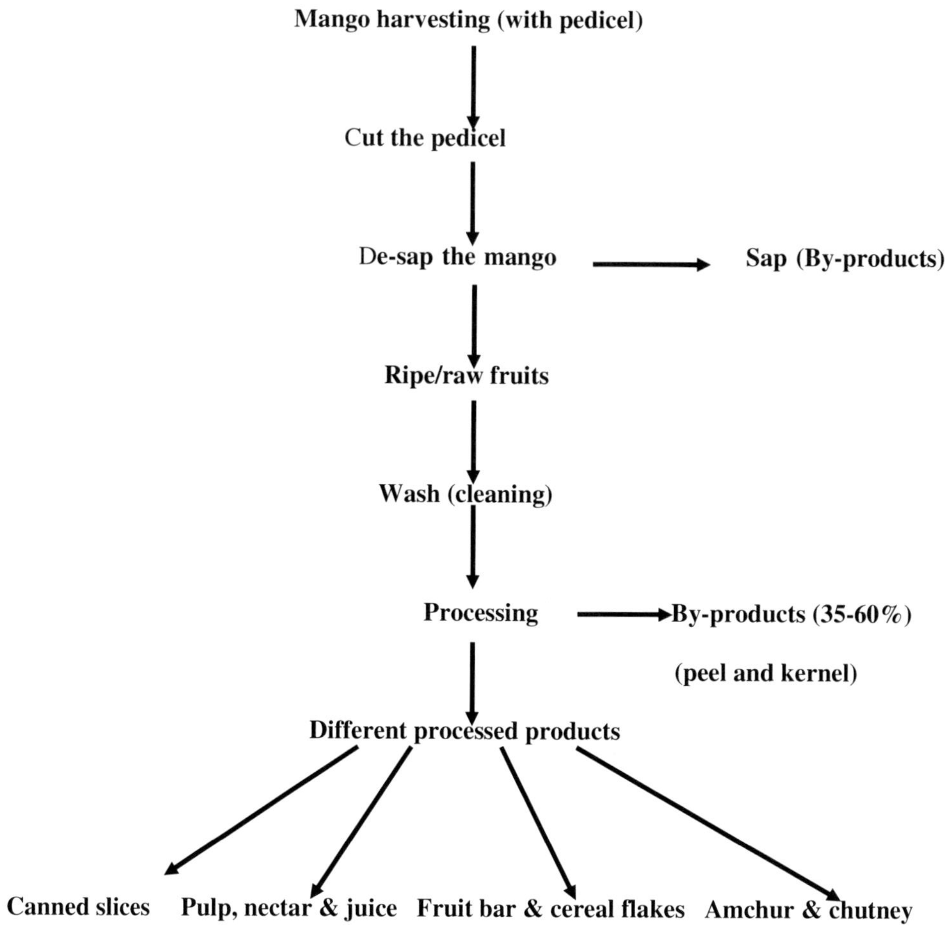

Figure 1. Processed foods and by-products generated during processing of mango

Mango Sap (Latex)

Latexes from different plants contain proteins, carbohydrates and polyphenols. Papaya latex contains papain, which is a well characterized protease, also comprising peroxidase and peptidase A [24,25]. Mango fruit has duct system, which continues into the pedicel (stalk) but ends before the abscission zone. Sap, which is a viscous, milky white liquid, is present in the duct system. During harvesting of mango fruit, when the pedicel is cut at the abscission zone, sap initially spurts out onto surface of the fruits and subsequently it oozes and causes sap-injury. To control sap-injury, one of the methods widely followed is to harvest the mangoes with pedicel, cut the pedicel at the abscission zone and de-sap the mango. Currently, sap collected is discarded and therefore, can be used as a by-product.

Sap can be separated into upper organic phase and lower aqueous phase by simple centrifugation [26, 27]. Aqueous phase of mango sap contains different enzymes, such as polyphenol oxidases, peroxidase and proteases, amylase and pectinase [11]. Among these enzymes, specific activities of peroxidase and polyphenol oxidases were reported to be very high compared to other enzymes. The activities of these oxidases varied in saps collected at different developmental stages of fruit. It was reported that the specific activities of peroxidases and polyphenol oxidases decreased in sap with increase in maturity of fruit and it varied from as low as 12% to 54% depending on the enzyme and mango variety. Isolation of different enzymes from mango sap by different researchers is given in Table 1. Recently, Saby John et al. [12] have purified and characterized the phenol oxidases viz., peroxidase, laccase and catechol oxidase from sap by a single step procedure of ion-exchange chromatography on DEAE-Sephacel column. The molecular weight of these enzymes was around 105,000 with a pH optima of 6 and the temperature stability varied between 50-70°C.

Table 1. Different enzymes in mango sap

Enzymes	References
Catechol oxidase	Saby John et al., [12]
Laccase	Joel and Fahn, [28]; Rabinson et al., [29]; Saby John et al., [11]
Peroxidase	Saby John et al., [11,12]
Protease	Saby John et al., [11]

Mango Peel

As mango is a seasonal fruit, about 20% of fruits are processed for products, such as puree, nectar, leather, pickles, canned slices and chutney and these products experience worldwide popularity and have also gained increased importance in the US and European markets [30]. Peel is one of the by-products obtained during mango processing and it constitutes about 15–20% of the fruit [31] and schematic flow diagram indicating the by-products generated during mango processing is been given in Figure 1. As peel is not currently utilized for any commercial purposes, it is discarded as a waste and becoming a source of pollution. Recently, Ajila et al. [3, 32] reported that peel is a good source of major nutrients, enzymes, polyphenols and carotenoids, and the peel extract had potential antioxidant properties (Table 2). As peel is an edible part of the fruit, it can be incorporated in foods to improve their nutraceutical properties.

Table 2. Proximate composition and antioxidant compounds of mango peel*

Component	Content
Moisture (%)	10.5
Fat (%)	2.2
Ash (%)	3.0
Protein (%)	3.6
Carbohydrate (%)	80.7
Total polyphenols (mg GAE/g MPP)	96.2
Carotenoids (μg/g MPP)	3092
Free radical scavenging activity, IC_{50} (μg MPP)	79.6

*Ajila et al., [32]; GAE- gallic acid equivalents; MPP- mango peel powder.

Table 3. Different enzymes in mango peel

Enzymes	References
Polyphenol oxidase	Prabha & Patwardhan, [33]; Saby John et al., [34]; Ajila et al.,[3]
Peroxidase	Saby John et al., [34]; Ajila et al., [3]
Protease	Ajila et al., [3]; Mehrnoush et al., [8-10]
Xylanase and amylase	Ajila et al., [3]

As shown in Table 3, mango peel also contains various enzymes like protease, polyphenol oxidase, peroxidase, xylanase and amylase [3,33,34]. Ajila et al. [3] determined these enzyme activities in two Indian mango varieties viz., Badami and Raspuri at different stages of maturity and reported that these enzyme activities in the mango peel varied depending on the variety and stages of its maturity. PPO activity was reported to be higher in ripe mango fruit peels compared to raw fruits [3,33,35].

Proteases are commercially important enzymes which have applications in various industrial sectors [36,37]. They are used for production of protein hydrolysates and tenderization of meat. Mango peel is rich in serine proteases [3] and recently, Mehrnoush et al. [8-10] optimized various conditions to obtain protease with high specific activity and stability. Using response surface methodology, Mehrnoush et al. [8] reported that optimum extraction conditions for serine protease from 100 g of mango peel were 50 ml of buffer, 2.5°C of temperature and 2 min of mixing in a Waring commercial laboratory blender. Mehrnoush et al. [10] also optimized conditions to purify serine protease from mango peel using polyethylene glycol (PEG)/dextran-based aqueous two-phase system (ATPS). In order to improve the storage stability of peel protease, it was encapsulated with gum arabic, maltodextrin and calcium chloride during freeze drying [9].

Mango peel and sap, the by-products of mango processing industry, could be a rich source of bioactive compounds and enzymes. This new source will be potential as a functional food or value-added ingredients in future with different applications. Mango by-products if conveniently processed, could furnish useful products that may balance out waste treatment costs and also decrease the cost of main product. Therefore, there is a scope for the isolation of these active ingredients especially enzymes and other value-added ingredients in food and enzyme industry.

Enzymes from Legume Processing Industry by-products

Legume seeds are sources of energy, fibre, amino acids, minerals, vitamins and essential fatty acids. The whole seed of most of the legumes is a rich source of energy, fibre, amino acids, minerals, vitamins and essential fatty acids. The legume seeds are used for human and animal consumption after minimal processing or for the production of oils. Among the plants, legumes are a valuable source of proteins. Soy bean, peanut (ground nut), black gram, green gram (mung bean), toor dhal (yellow pigeon pea), cowpea, horse gram and bengal gram (chick pea) are some of the major legumes grown in the world and they belong to Leguminacae family. Most of these legumes are also called pulses. The legume grain containing low oil (1-5%), high protein (20-30%) and carbohydrate (50% or more) are called pulses. India is the world's largest producer (about 26%) and consumer of pulses. In India, pulses are the second major sources of dietary proteins after cereals.

Pulses are relatively inexpensive sources of dietary proteins compared to animal proteins and possess desirable attributes, such as abundance of complex carbohydrates, ability to lower cholesterol in humans, low fat content, higher concentration of polyunsaturated fatty acids, long shelf-life and diversity of foods that can be made from them. They also contain bioactive compounds whose beneficial effects in human health need to be fully exploited [38]. Pulses are found to be rich in oxidative enzymes, such as peroxidase, which has many physiological roles, and industrial and analytical applications.

Basic staple diets, such as rice, wheat and legumes undergo primary mechanical processing operations that convert the grain into edible raw material. After primary processing, the products obtained are subjected for secondary processing to yield popular products, such as bread, biscuits and many other food products. During primary processing, the major by-products obtained are seed coat and germ. The milling of legumes is performed to transform grains into dhal (split cotyledon). About 75% of total pulse produced in India is milled into dhal. The major operations involved in legume processing are cleaning, drying, milling, packaging, handling and storage. Legume milling basically consists of removal of the pericarp seed coat, which is attached to the cotyledon through a layer of biochemically complex gum. Pre-milling treatment involves loosening the pericarp from cotyledons through chemical, mechanical or heat treatment. Milling of legumes essentially consists of the removal of the tightly bound external 'husk' from the grain and recovery of the cotyledon in the form of dhal. The average yield of dhal in the commercial mill is about 75% and the rest of grain material (about 25%) is obtained in the form of various types of by-products. These comprise of husk, powder, large and small brokens, shriveled and under processed grains [39].

The legume by-products were found to be a rich source of valuable compounds such as antioxidants, dietary fibers and enzymes [40, 41]. The extraction of antioxidant compounds from cereal and legume by-products such as seed coat, germ etc have been reported [5,42]. Seed coat, which is also called hull, or husk, serves as a multifunctional organ in legume seeds. It transports, transforms and secretes various metabolites to the embryo [43], and to the mature seed it provides structural protection and mechanical support. Due to the presence of various plant defence compounds, seed coat protects the seed against pathogens, insects, wind, frost, and other stresses [44]. The seed coat of cereals and legumes contain large quantities of endogenous antioxidants such as phenolic compounds, which exhibits antioxidant and antimicrobial properties [5,42, 45]. Soy bean seed coat was also reported to contain plant defense proteins and peroxidase [46]. Legume by-products such as soybean hull and black gram husk are found to be a rich source of enzymes such as peroxidase [41,47].

Soybean by-products

The soybean (*Glycine max*) is often called the miracle crop. It is the world's foremost provider of protein and oil. Soybean is the largest oilseed crop in the world accounting for more than 50% of the world oilseeds production. In 2011, soybeans represented 56% of world oilseed production, and 33% of those soybeans were produced in the United States followed by Brazil (25%), Argentina (19%), China, India, Paraguay and Canada (http://www.

soystats.com/2012). Soybean is processed into various products such as soybean oil, soy flour, soy milk, soy milk powder, tofu, soy sauce etc. Most of the soybeans are used in the production of soybean oil. During the process, soybean byproducts such as stalks, leaves and husk are produced. Soybean by-products, especially soybean hull are normally used as animal feed [48]. Soybean hulls with an US production more than $9.1X10^8$ kg are an underutilized co-product of the soybean processing industry [41]. Soybean seed coats, or hulls, represent about 8 to 10% of the weight of soybean grain [49]. The nutrient composition of soybean hull is given in the Table 4. Soybean hulls, though high in fiber, are a significant source of protein, comparable to corn grain in the amount of crude protein.

Table 4. Composition (%) of soybean hull*

Component	Content (g/100g)
Cellulose	14 - 25
Hemicellulose	14 - 20
Pectin	10 - 12
Protein	9 - 12
Uronic acid	7 - 11
Ash	4 – 5
Lignin	3 - 4

* Mullin and Xu [40].

Table 5. Protein and enzymes identified from soybean seed hull

Protein/Enzymes	Function	References for source and functions
Proline-rich proteins	----	Percy et al., [55]
Kunitz & Bowman–Birk trypsin inhibitors	chemopreventative & anticarcinogenic agent	Gijzen et al., [54]; Sessa and Wolf, [49]
Peroxidase	Various physiological functions in plants including disease resistance, wound healing; analytical biochemistry and water purifications and pesticide degradation, biosensors,	Gijzen et al., [52]; Gijzen, [56]; Gillikin and Graham, [53]; Smith and Hammerschmidt, [57]; Spelie et al., [58]; Hammerschmidt et al.,[59]; Sessa and wolf, [49]; Yoshida et al., [60]; Hamid and Khalil-ur-Rehman, [61]
Chitinase	protecting the seed from fungal attack; chitooligosaccharide production	Gijzen et al., [54]; Patil et al., [62]; Wang and Hwang, [63]
Hydrophobic protein	----	Gijzen et al., [64]

At present soy hulls are primarily used as fiber sources for dairy and beef cattle. Soy hull proteins have been investigated for almost three decades and the type of proteins/enzymes reported in Table 5. Soybean hulls are low in soluble fiber, and have only modest and inconsistent effects on serum cholesterol levels [50]. Soybean Bowman-Birk trypsin inhibitor is an anti-cancer agent that can be obtained from seed coats of soybean cultivar Williams [49]. Soybean grain has been identified as a good source of available iron [51]. Soybean husk was found to be a rich source of peroxidase enzyme [49,52]. Soybean hulls of some cultivars have a high peroxidase activity, 100-fold greater than that of cultivars containing low activity. The difference in activity is conditioned by a single gene pair, designated Ep/ep [52]. Gillikin and Graham [53] purified peroxidise from soy bean seed coat to homogeneity by successive chromatography on DEAE Sepharose Fast Flow, concanavalin A-Sepharose, and Sephadex G-75. Some of the characteristic properties of the peroxidase were that its molecular weight was 37 kDa, isoelecticc point was 4.1 and it was as a glycoprotein. Soybean hulls also showed chitinase activity which was expressed later in seed development. This was believed to serve a defensive function, perhaps in protecting the seed from fungal attack [54].

Black Gram Byproducts

Black gram or urad (*Phaseolus mungo* L.) is one of the important pulse crops in India. The production of pulses during 2005-06 estimated at 10.6 million tonnes, while black gram production was around 1.1 million tonnes [65]. Protein content in black gram ranges from 26-28% (dry-weight basis), which is almost three times that of cereals. Globulin is the major storage protein, which accounts for 81% of the total protein, followed by albumin, prolamine and glutelins. They are deficient in sulphur-containing amino acid and rich in lysine [66]. It is consumed mainly in the form of dhal (split cotyledons).

About 75% of total black gram produced in India is milled into dhal. It is processed at domestic, cottage and small-medium scale industries. During milling of black gram into dhal, significant amount of waste by-products are obtained. These comprise of husk, powder, large and small brokens, shriveled and under processed grains [39]. Ajila and Prasada Rao [47] reported that around 25% is the by-product obtained during milling of black gram into dhal and they separated them into husk, germ, plumule and aleurone rich husk rich fractions using air classification methods (Figure 2). The proximate composition of black gram husk on dry-weight basis has been reported as shown in Table 6. The husk of black gram was investigated as a biosorbent of cadmium from low concentration aqueous solutions.

Table 6. Proximate composition (%) of black gram husk*

Component	Content
Moisture	15.5
Crude protein	3.32
Fat	0.7
Carbohydrates	39.15
Crude fiber	51.9
Ash	4.93

* Saeed & Iqbal [67].

The black gram milled by-product is currently used for cattle feed or wasted and, therefore, it does not have any commercial value. However, these are rich in various valuable compounds including enzymes and bioactive compounds [5,47]. Compared to pure husk fraction and germ, aleurone rich husk was shown to be a good source of enzymes mainly peroxidase enzyme [47]. Black gram husk peroxidase had pH optimum at 5.5 and thermal stability up to $70^{\circ}C$. Its molecular weight was found to be 35kDa. As the enzyme showed wide substrate specificity it may belong to class III of the plant peroxidase superfamily similar to horseradish peroxidase or soy hull peroxidase.

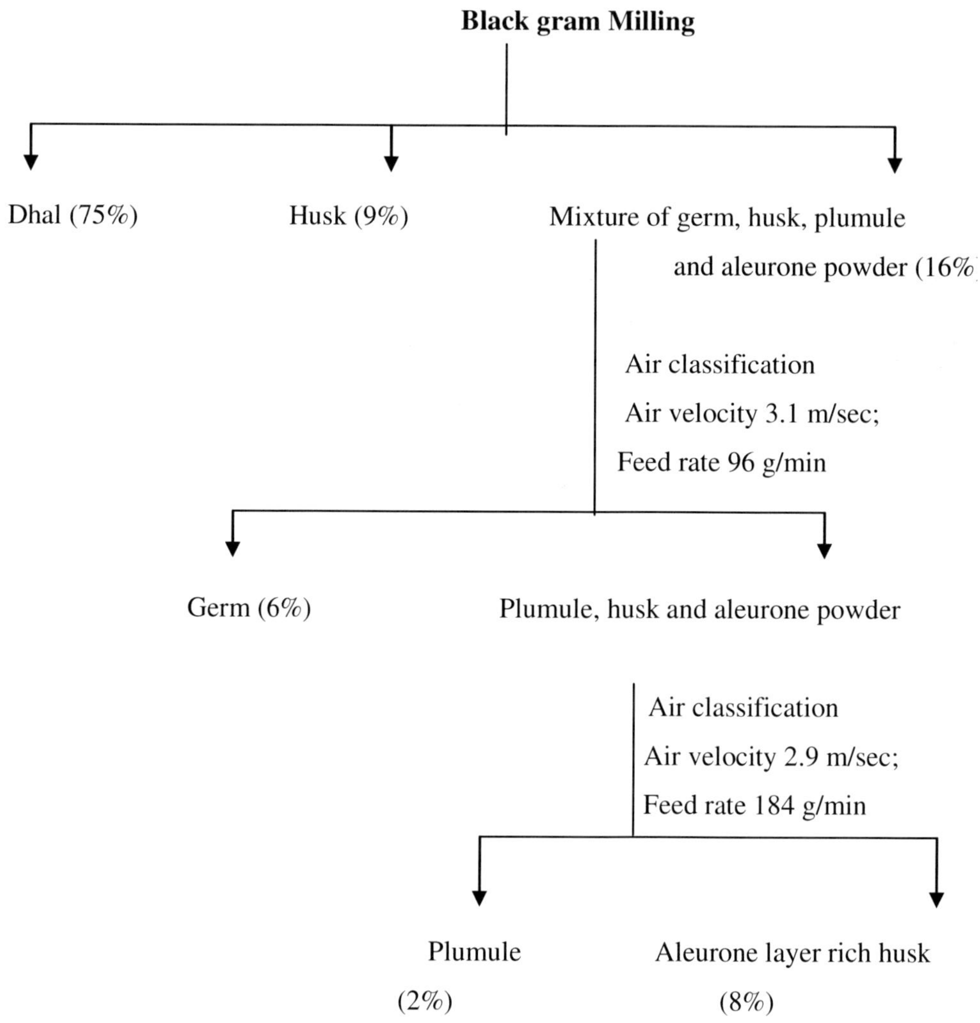

Figure 2. Scheme for separation of black gram milled by-products (Source ref. 47).

Utilization of waste material for production of valuable compounds boosts up higher economic returns in many industries. Advancement in biotechnology improved the potential use of agro-industrial residues for value addition by production of enzymes, organic acids, bioactive secondary metabolites and single-cell protein, among others. Legume by-products which are rich in nutrients and other valuable compounds can be used for the recovery of many valuable compounds as described above.

PRODUCTION OF ENZYMES FROM MANGO AND LEGUME BYPRODUCTS BY FERMENTATION

Food processing industry by-products can be used in a number of ways especially for biomass production. Various types of by-products can be used to grow microorganisms to produce enzymes, single cell protein, amino acids, lipids, carbohydrates and organic acids, which have animal feed and food processing applications. Application of agro-industrial wastes in bioprocesses provides an alternative way to replace the refined and costly raw materials. In addition, the bulk use of such materials helps to solve many environmental hazards. Protein enrichment has been obtained by growing *Aspergillus niger* on mango peel, orange peel, green immature banana and carrot wastes in solid state fermentation. Renewable energy sources, such as ethanol, methane and bio-hydrogen also can be produced by fermentation of agro industrial by-products. Bioconversion of renewable cellulosic biomass into fuel ethanol as an alternate to petroleum is rising around the world [68]. In developing countries, production of biogas from waste is an appropriate solution. Biogas generation through anaerobic fermentation is considered to be a simple and economical system of waste treatment and a number of processes have been developed and commercially exploited. Anaerobic digestion of mango peel resulted in biogas production of 0.33 m^3/kg with 53% methane content [69]. This section describes briefly about the utilization of mango processing byproducts and legume by-products for the production of enzymes through fermentation techniques.

Utilization of Mango by-products

Production of value-added products from agro-industrial and food processing wastes is now a focusing area, as it reduces pollution in the environment in addition to energy generation. Majority of fruit and vegetable wastes available from their processing industries are seasonal and they do not decompose rapidly. The mechanical drying of these wastes (mango peel, citrus peel, pineapple peel and tomato processing wastes) gave opportunity to store the substrate all over the year. Mango is processed to a maximum extent, thereby producing high quality of solid and liquid wastes. Solid wastes, stones, stalks, trimmings and fibrous materials are obtained during the preparation of raw material which contributes about 40 to 50% of total fruit waste out of which, 5 to 10% is pulp waste and 15 to 20% is kernel [70]. Liquid waste is the waste material that comes out of a factory after washing of fruits, packaging, blanching, cooling and plant and machinery clean up and so on. Utilization of this mango waste is both a necessity and a challenge. If a factory is processing five tons of Totapuri mangoes per hour, about six tons of peel would be available as waste per day of 8 h work. Approximately, 0.4 to 0.6 million tons of mango peel is generated annually in India [71]. This waste is either used as cattle feed or dumped in open areas, where it adds to environmental pollution. Mango peel is difficult to decompose, as it takes a very long time, because of its complex composition. The presence of high amount of reducing sugars in dried and fresh mango peel prompted to utilize it as a raw material for enzyme and ethanol production and development of cheap medium. Production of different enzymes using mango processing industry by fermentation is summarized in Table 7.

Plant cell degrading enzymes are used industrially for enhancing the quality of products and improving the efficiency of the clarification of fruit juices and wines as well as the extraction of oils. Enzyme preparations for these usually contain the mixture of hydrolytic enzymes such as polygalacturonase, cellulase, FPCAse and hemicellulase. A mixture of enzymes containing polygalacturonase, cellulase, xylanase and protease enzymes was produced using *Aspergillus niger* 3T5B8 using different agro-industrial residues such as banana peel, mango peel, cassava bran and wheat bran by solid-state fermentation and tested for vegetable oil extraction [72]. It was also reported that mango peel was the best substrate for FPCase production, mainly in the presence of 0.2% cellobiose [72].

Pectinases are group of enzymes that catalyze the breakdown of pectin containing substances and can be produced either by submerged fermentation or solid state fermentation methods. Pectinolytic enzymes have been classified on the basis of their action on the α-1,4-glycosodic bond of methyl esterified D-polygalacturonic acid. The depolymerizing enzymes split the glycosidic bonds of pectic acid or pectin by hydrolysis (hydrolases) or by β-elimination (lyases), while pectin methylesterase de-esterifies the pectin without changing the degree of polymerization. These enzymes are widely used for industrial applications such as extraction of the juice from fruits, juice clarification, coffee and tea fermentations, extraction of essential oils and wine production [73]. Pectinases produced by different microbes are divided into depolymerizing enzymes and saponifying enzymes. Depolymerizing enzymes are polymethylgalacturonases, pectin lyases, polygalacturonases and pectate lyases and saponifying enzymes are pectinesterases [74]. Fungal polygalacturonases are very important for clarification of fruit juices, wines and for extraction of vegetable oils and their significance in clarification of fruit juices is due to the fact that their optimal pH closer to that of many fruit juices [75]. Use of fruit wastes, such as jack fruit rind and mango peel as substrates for production of polygalacturonases by *Aspergillus awamori* MTCC 9166 in solid state fermentation was reported and it was found that the enzyme production by this strain was higher than the reported strains [75]. Evaluation of mango peel as a substrate for the production of pectinase enzyme by using *Aspergillus foetidus* was done in solid state and submerged fermentation systems and the highest productivity of polygalacturonase and pectin lyase was obtained with solid state fermentation and is less affected by catabolic repression [76].

Polygalacturonases are produced by numerous fungi and bacteria and also by higher plants [77]. These plant enzymes are produced prior to budding or branch formation to break down existing cell wall. Polygalacturonases (PGases) catalyze the random hydrolysis of 1,4 α-D galacturonic acid linkages in smooth region of pectin. The pectinolytic enzymes degrade the pectic substances found in plant tissues, thereby, having numerous applications in various types of industries, such as juice and food industries, paper and pulp industry and natural fibres treatment in textile industries [78]. Feasibility of agro-industrial residues for pectin lyase and polygalacturonase production by a newly isolated thermophilic *Thermoascus aurantiacus* was reported by Martins et al. [79]. This fungus is able to produce high levels of extracellular enzymes during the SSF of mixtures of sugar cane and orange bagasses. Agro-industrial wastes, such as wheat bran, cassava, sugar beet pulp, citrus waste, corn cob, banana waste, saw dust and fruit pomace are the most commonly used substrates for SSF for pectinase production [70]. Recently, Patil and Dayanand [62] have evaluated pectin-rich agro-wastes, viz. lemon peel, sorghum stem and sunflower head, as substrates for the production of pectinase by *Aspergillus niger* DMF 27 and *A. niger* DMF 45 in SMF and SSF systems, respectively. The increased level in the production of pectinases was detected when the agro-wastes were supplemented with additional carbon and nitrogen sources, the addition of sucrose being more effective than glucose in SSF. Other comparative reports have been found that SSF was more productive than SMF and, in addition, the pectinases produced by SSF showed more stable properties: they had a higher stability to pH and temperature and they were less affected by catabolic repression than pectinases produced by SMF [80,81]. Production of polygalacturonase from mango peel and optimization of all processes has been reported using Taguchi methodology by *Fusarium moniliforme* NCIM 1276 under SSF conditions [82]. Optimization of the enzyme production clearly demonstrated the impact of the process parameters on the gross yield of the polygalacturonase enzyme. The optimized medium and conditions composed of temperature (30 C), mango peel (6.5%, g, w/v), inoculum (8%, ml, v/v), peptone (1%, g, w/v), NH_4NO_3 (0.60%, g, w/v), $MgSO_4$ (0.05%, g, w/v), $ZnSO_4$ (0.06%, g, w/v) and KH_2PO_4 (0.4%, g, w/v). At this optimum condition, the yield of polygalacturonase production by *F. moniliforme* was found to be 43.2 U/g [82].

Table 7. Production of different enzymes from mango processing industry by fermentation

Byproducts	Enzymes	Micro-organisms	References
Banana peel, mango peel, cassava bran & wheat bran	Mixture of polygalacturonase, cellulase, xylanase & protease enzymes	*Aspergillus niger*	Couri et al., [72]
Jack fruit rind and mango peel	Polygalacturonases	*Aspergillus awamori*	Padma et al., [75]
Mango peel	Pectinase	*Aspergillus foetidus*	Kumar et al., [76]
Mango peel	Polygalacturonase	Fusarium moniliforme	Kumar et al., [82]
Mango peel	Carboxymethyl cellulase	*Paenibacillus polymyxa*	Devendra kumar et al., [83]

Carboxymethyl cellulase (CMCase), a major constituent of cellulase complex is widely used in chemicals, fuel, food, animal feed, brewery and wine, textile and laundry, pulp and paper and agro-based industries [84]. Since, mango peel is a rich source of lingocellulosic fibre, its potential use as substrate for CMCase production under optimized conditions was studied. The study revealed that mango peel may serve as substrate for CMCase production using *P. polymyxa* with optimized conditions. Maximum CMCase production (7.814 U mg-1) was observed in a medium containing 7% mango peel (w/v) with 1.5% ammonium sulphate (w/v) at 37°C and pH 5.5 [83].

Utilization of Legume by-products

Food legumes are major source for protein in our diet. Most legumes are consumed after a simple industrial process, in which the pod is removed to prepare the seed. Peas and broad beans are widely consumed, and are grown all over the world. The large quantities of by-products generated during the processing of plant food involve an economic and environmental problem due to their high volumes and elimination costs and now they are considered as a promising source of enzymes and bioactive compounds [85]. After isolation of the main constituent, these byproducts represent an inexpensive material that has been undervalued until now, when it was only used as fuel or fertilizer. With the development in biotechnology, attempts have increasingly been made globally to make potential use of agro-industrial residues for value-addition by production of enzymes, organic acids, bioactive secondary metabolites, single-cell protein, among others. Lot of research has been carried out on production of valuable enzymes from many legume processing by-products by employing fermentation techniques using suitable microorganisms. Table 8 provides information on the production of different kind of enzymes by fermentation of legume by-products.

Tannase is an important enzyme in many industries, such as in the manufacture of instant tea, beer, fruit juice, coffee-flavour, soft drink, grape wine and pharmaceuticals. Tannase catalyses the breakdown of hydrolysable tannins, such as tannic acid, methylgallate, ethyl gallate, n-propylgallate and isoamylgallate. A typical reaction of tannase is the hydrolytic cleavage of (-)epigallocatechin-3-ol gallate [86]. There has been a report on the utilization of red gram husk as a substrate for tannase production and optimization of the temperature and incubation period of tannase production in solid-state fermentation using *Aspergillus niger*. The maximum production of tannase was found to be 43 U/g/min at 35°C, incubated for 96 h. The tannin content in the substrate was decreased during the incubation period [87]. Red gram ranks sixth among the production of pulses in the world and it is the major legume crop. In India it is the second largest pulse crop accounting about 20 percent of total pulse production and this legume is processed into different food products with generation of wastes, such as husk. India annually produces about 2.0-2.5 million tonnes of red gram and the production has been stagnant in the past 10 years [87].

Proteases are large group of enzymes and also known as proteinases or proteolytic (protein-digesting) enzymes, which are mainly classified on the basis of their pH optimum as acidic, neutral, and alkaline proteases. Proteases are involved in digesting long protein chains into short fragments, splitting the peptide bonds that link the amino acid residues. Most of the enzymes employed in industrial production of foodstuffs, biochemical reagents and other important compounds originate mainly from bacteria, especially from varieties of genus *Bacillus* [88]. Microbial proteases can be produced from bacteria, fungi and yeast using many processes, such as solid-state and submerged fermentation [89]. Microbial protease represents about 60% of the worldwide sales of the industrial enzyme and it is due to their applications in several industrial sectors, such as detergent, food, pharmaceuticals, chemicals, leather, paper and pulp and silk industries. Among all the proteases, alkaline proteases are robust in nature and are primarily used as detergent additives [36]. They account for 40% of the total worldwide enzyme sales and this trend is expected to increase in near future [90]. At present, the overall cost of enzyme production is very high due to high cost of substrates and mediums used. Recently, efforts have been directed to explore the means to reduce the protease production costs through improving the yield, and the use of low-cost feed stocks or agricultural byproducts as substrates for protease production [91]. There are several reports describing the use of agro-industrial residues for the production of alkaline protease, e.g. nug meal and *Bacillus sp.* AR009 [92], pigeon pea and *Bacillus sp.* JB-99, wheat bran and *Rhizopus oryzae* [93]. Alkaline protease production by *Bacillus cereus* under solid-state fermentation of red gram husk was influenced by physiological and chemical nature of the red gram husk and associated with growth of the microbial strain [94]. Red gram husk supported good bacterial growth and enzyme production (protease, 262 U/g of substrate). Maximum protease production was observed at a growth of 48 h, a substrate pH of 9.0, 50 % substrate moisture, temperature of 40°C and 7.5g of substrate concentration. Enrichment with fructose as carbon source and beef extract as nitrogen source enhanced the enzyme yield reaching 258 U protease activities per gram of red gram husk. The enzyme production was found to be associated with the growth of the bacterial culture [94]. The choice of the nitrogen and carbon sources has a major influence on the yield of protease.

Production of alkaline protease from locally available, inexpensive agro-substrate, green gram husk using *Bacillus* sp. under solid-state fermentation was reported [95]. Alkaline protease production by isolated *Bacillus sp.* strain under solid-state fermentation was influenced by physiological and chemical nature of the gram husk and associated with growth of the microbial strain. Particle size of the solid material regulated the enzyme production and yield was improved with the supplementation of carbon and nitrogen sources to the solid medium. Optimum enzyme production was achieved with 1.5% maltose and 2.0% yeast extract with 371% increase than the control. Glucose did not repress the enzyme production but inorganic nitrogen sources showed little negative impact. The physiological fermentation factors, such as pH of the medium (pH 9.0), moisture content (140%), incubation time (60 h) and inoculum level played a vital role in alkaline protease production. The enzyme production was found to be associated with the growth of the bacterial culture [95].

Table 8. Production of different kind of enzymes by fermentation of legume by-products

Byproducts	Enzymes	Micro-organisms	References
Red gram husk	Tannase	*Aspergillus niger*	Paranthaman et al., [87]
Red gram husk	Alkaline protease	*Bacillus cereus*	Rathakrishnan and Nagarajan, [94]
green gram husk	Alkaline protease	*Bacillus sp*	Prakasham, [95]
Bengal gram husk	L-Glutaminase	*Bacillus sp*	Sathish et al., [96].
Wheat bran, rice bran and green gram husk	Amylase	*Aspergillus sp.*MK07	Chimata et al., [97]
Black gram	Protease	Bacillus subtilis	Meruvu & Vangalapati., [98]

L-Glutaminase is an important enzyme in pharmaceutical industry and has a pivotal role in cellular nitrogen metabolism [99]. This enzyme is important in industrial and pharmaceutical sectors as an effective agent in the treatment of acute lymphocytic leukemia [100] and HIV [101], as an analytical agent [102], as a biosensing agent [103], as a flavour-enhancing agent [104] and in the production of specialty chemicals, such as threonine by gamma glutamyl transfer reactions [105]. Several studies reported that the use of agro-industrial waste materials improved the microbial product production with the supplementation of agro-material specific deficient nutrient sources [95,106]. Glutaminase production has been reported under solid-state fermentation using wheat bran, sesame oilcake and polystyrene beads [107,108]. Bengal gram husk found to be one of the most suitable substrate medium for glutaminase production by *Bacillus sp.* Maximum glutaminase production is achieved using solid-substrate mixture at two-level combinations in the ratio of 66: 34 for Bengal gram husk and wheat bran, respectively [96].

Amylases are a group of important enzymes which are mainly used in starch processing industries for the hydrolysis of polysaccharides into simple sugars [109]. Amylases accounts for about 30% of the world's enzyme production [110]. Amylase has various applications in sectors, such as confectionary, baking, paper, textile, detergent and pharmaceutical etc [111]. Amylases are produced on a large scale normally by solid state fermentation (SSF) mainly due to advantages, such as lower capital expenditure, cheaper fermentation media, superior productivity, reduced energy, requirements and absence of rigorous control of fermentation parameters. Production of extracellular amylase by solid-state fermentation was investigated using *Aspergillus sp.*MK07 from agricultural residual substrates, such as wheat bran, rice bran and green gram husk, among others. Effects of various process variables, such as incubation period, temperature, initial moisture content, pH, supplementary carbon, nitrogen source and inoculum level on production of amylase have been studied. It was found that amylase production was highest at 120 h of incubation period at 30°C, 70% initial moisture content, 5.0 pH and 5% inoculum level. Supplementation of carbon (starch) and nitrogen source (peptone) showed an increase in amylase production and the higher amount of amylase production obtained under all optimized conditions was 164 U/g [97].

Enzymes are one of the most important biological catalysts for many industrial and medical applications. Global enzyme demand increases yearly based on the strong demand in the specialty enzyme segment and growth in different industrial applications. The cost and availability of substrates also play an important role in the enzyme production. The utilization of agro residues for the commercial production of enzymes has been well explored. Agro-industrial by-products can be successfully utilized for solid-state fermentation to produce the microbial production enzymes due to their low cost and availability.

CONCLUSION

Global concept of utilization of by-products is changing rapidly in response to need for environmental conservation, sustainable agricultural productivity and global food security. Biotechnological processes, such as solid substrate fermentation and novel enzyme purification technologies offer immense opportunities for the use of abundant agricultural wastes for the production of valuable compounds including enzymes. Enzymes are one of the most important biological catalysts for many industrial and medical applications. The utilization of agro-residues for the commercial production of enzymes has great potential. Agro industrial by-products can be successfully utilized for solid-state fermentation to produce the microbial production enzymes due to their low cost and availability.

REFERENCES

[1] H.A. Qdais, F. Abdulla and L. Qrenawi, *Jordan J. Mechanical Industrial Eng.* 4, 69 (2010).
[2] A. Buekens and H. Huang, *J. Hazard Materials,* 62, 1 (1998).
[3] C.M. Ajila, S.G. Bhat, and U.J.S. Prasada Rao, *Food Chem.* 102, 1006 (2007a).
[4] B.T. Manu and U.J.S. Prasada Rao, *Food Chem.* 114, 66 (2009).
[5] T.K. Girish, V.M. Pratape and U.J.S. Prasada Rao, *Food Res International,* 46, 370 (2012).
[6] D. P. Makris, G. Boskou and N.K. Andrikopoulos, *Bioresource Technol.* 98, 2963 (2007).
[7] S. Ketnawaa, P. Chaiwutb and S. Rawdkuena, *Food Bioproducts Processing,* 90, 385 (2012).
[8] A. Mehrnoush, P.T. Chin, M. Hamed, A.A. Norashikin. C.L. Tau, *Food Chem.* 124, 666 (2010).
[9] A. Mehrnoush, C.P. Tan, H. Mirhosseini, A.A. Norashikin and C.L. Tau, *Food Chem.* 128, 158 (2011).
[10] A. Mehrnoush, S. Mustafa, M.Z.I. Sarker and A.M.M. Yazid, *International J. Mol. Sci.* 13, 3636 (2012).
[11] K. Saby John, S.G. Bhat, and U.J.S. Prasada Rao, *Phytochemistry,* 62, 13 (2003).
[12] K. Saby John, S.G. Bhat, and U.J.S. Prasada Rao, *J. Molecular Catal B: Enzymatic,* 68, 30 (2011).
[13] M.A. Milala, A. Shugaba, A. Gidado, A.C. Ene and J.A. Wafar, *J. Agric. Biol. Sci.* 1, 325 (2005).
[14] C.F. Gonzalez, J.I. Farina and L.I.C. de Figueroa, Enzyme Microbiol. Technol. 42, 272 (2008).
[15] V. Arasaratnam, K. Mylvaganam, and K. Balasubramaniam, *J. Food Sci. Technol.* 38, 334 (2001).
[16] C. Botella, A. Diaz, I. Ory, C. Webb, and A. Blandino, *Process Biochem.* 42, 98 (2007).
[17] D. Chapla, J. Divecha, D. Madamwar and A. Shah, *Biochem. Eng. J.* 49, 361 (2010).
[18] R.A. Bayoumi, H.M. Yassin, M.A. Swelim and E.Z. Abdel-Al, *J. Appl. Sci. Research,* 4, 1708 (2008).
[19] P.T. Sangeetha, M.N. Ramesh and S.G. Prapulla, *Process Biochem.* 39, 755 (2004).
[20] V. Pradeep and M. Datta, Appl. Biochem. Biotechnol. 102, 109 (2002).
[21] K.N. Niladevi, R.K. Sukumaran, and P. Prema, *J. Indian Microbiol. Biotechnol.* 34, 665 (2007).
[22] N. Vassilev, A.R. Requena, L.M. Nieto, L. Nikolaeva and N. Vassileva, Ind. Crop Prod. 30, 28 (2009).
[23] FAO Statistics, Food and Agriculture Organization of the United Nations, Rome, Italy. http://faostat.fao.org/. (2007).
[24] K. Brocklehurst, E. Salih, R. McKee and H. Smith, *Biochemical J.* 228, 525, (1985).
[25] L-C. Chen, Y-C. Chung, -Chin, and C-T. Chang, *Food Chem.* 135, 2529 (2012).
[26] B.R. Loveys, S.P. Robinson, J.J. Brophy and E.K. Chacko, *Austr. J. Plant Physiol.* 19, 449 (1992).

[27] K. Saby John, L. Jagan Mohan Rao, S.G. Bhat and U.J.S. Prasada Rao, *Phytochemistry,* 52, 891(1999).
[28] D. M. Joel and A. Fahn, *Annals Botany,* 46, 785 (1980).
[29] S.P. Robinson, B.R. Loveys and E.K. Chacko, *J. Plant Physiol.* 20, 99 (1993).
[30] L. Loelillet. *Fruit,* 49, 332 (1994).
[31] O.P. Beerh, B. Raghuramaiah, G.V. Krishnamurthy, and N. Giridhar, *J. Food Sci. Technol.* 13, 138 (1976).
[32] C.M. Ajila, K.A. Naidu, S.G. Bhat, and U.J.S. Prasada Rao, *Food Chem.* 105, 982 (2007b).
[33] T.N. Prabha and M.V. Patwardhan, *Acta Alimentaria,* 15, 123. (1986).
[34] K. Saby John, S.G. Bhat, and U.J.S. Prasada Rao, *J. Food Biochem.* 26, 403 (2002).
[35] M. Ueda, K. Sasaki, N. Utsunomiya, K. Inaba, and Y. Shimabayashi, *Food Sci. Technol. Res.* 6, 299 (2000).
[36] R. Gupta, Q. Beg, and P. Lorenz, *Appl. Microbiol. Biotechnol.* 59, 15 (2002).
[37] A. Sumantha, P. Deepa, C. Sandhya, G. Szakacs, C.R. Soccol and A. Pandey, *Brazilan Arch. Biol. Technol.* 49, 843 (2006).
[38] C.W. Beninger, and G.L. Hosfield, (2003). *J. Agric. Food Chem.* 51, 7879-7883.
[39] N. Ramakrishnaiah, V.M. Pratape, V.B. Sashikala and H.V. Narasimaha, (2004). *Journal of Food Science and Technology, 41,* 492-496.
[40] W.J. Mullin and W. Xu, *J. Agric. Food Chem.* 49, 5331(2001).
[41] Sessa, D.J. (2004). Processing of soybean hulls to enhance the distribution and extraction of value-added proteins. *Journal of Science of Food and Agriculture, 84*, 75–82.
[42] A. Moure, J. M. Cruz, D. Franco, J. M. Dominguez, J. Sineiro, H. Dominguez, M.J. Nunez, J. C. Parajo, *Food Chem.* 72, 145 (2001).
[43] J.W. Patrick and C.E. Offler, *J. Exp.Bot.* 52, 551 (2001).
[44] F.D. Boesewinkel, and F. Bouman, Seed development and germination, J. Kigeland G. Galili (eds). New York: Marcel Dekker, 1 (1995).
[45] T. Tsuda, M. Watanabe, K. Ohshima, S. Norinobu, S. Choi, S. Kawakishi and T. Osawa, *J. Agric. Food Chem.* 42, 2407 (1994).
[46] S. Dhaubhadel, K. Kuflu, M.C. Romero and M. Gijzen, *J. Exp. Bot.* 56, 2335 (2005).
[47] C.M. Ajila, and U.J.S. Prasada Rao, *J. Molecular Catal B: Enzymatic,* 60, 36 (2009).
[48] K.M. Chee, K.S. Chun, B.D. Huh, J.H. Choi, M.K. Chung, H.S. Lee, I.S. Shin and K.Y. Whang, *Asian-Australian J. Animal Sci.* 18, 861 (2005).
[49] D.J. Sessa and W.J. Wolf (2001). Bowman–Birk inhibitors in soybean seed coats. *Industrial Crops Production, 14*, 73–83.
[50] F.L. Shinnick, R. Mathews and S. Ink, *Cereal Foods World,* 36, 815 (1991).
[51] L.E. Murray-Kolb, R. Welch, E. C. Theil and J.L. Beard, *Am. J. Clin. Nutr.* 77, 180 (2003).
[52] M. Gijzen, R. Van Huystee, and R.I. Buzzell, *Plant Physiol.* 103, 1061 (1993).
[53] J.W. Gillikin and J.S. Graham, *Plant Physiol.* 96, 214 (1991).
[54] M. Gijzen, K. Kuflu, D. Qutob, and J.T. Chernys, *J. Exp. Bot.* 52, 2283 (2001).
[55] J.D. Percy, R. Philip and L.O. Vodkin, *Plant Mol. Biol.* 40, 603 (1999).
[56] M. Gijzen, *The Plant J.* 12, *991* (1997).
[57] J.A. Smith and R. Hammerschmidt, *Physiology Mol. Plant Pathol.* 33, 255-261. (1988).
[58] K.E. Spelie, V.R. Franceschi and P.E. Kolattukudy, *Plant Physiol.* 81, 487 (1986).

[59] R. Hammerschmidt, D.T.A. Lamport, and E.P.Muldoon, *Physiol. Plant Pathol.* 24, 43 (1984).

[60] K. Yoshida, P. Kaothien, T. Matsui, A. Kawaoka, and A. Shinmyo, *Appl. Microbiol. Biotechnol.* 60, 665 (2002).

[61] M. Hamid and Khalil-ur-Rehman. *Food Chem.* 115, 1177 (2009).

[62] S.R. Patil, and A. Dayanand, *Bioresource Technol.* 97, 2054 (2006).

[63] S.L. Wang and J.R. Hwang, *Enzyme. Microb. Technol.* 28, 376 (2000).

[64] M. Gijzen, S.S. Miller, K. Kuflu, R.I. Buzzell, and B.L.A. Miki, *Plant Physiol.* 120, 951 (1999).

[65] India Infoline (2005).

[66] N.R. Reddy, D.K. Salunkhe and S.K. Sathe, *CRC Critical Rev. Food Sci. Nutr.* 15, 49 (1982).

[67] A. Saeed and M. Iqbal, *Water Res.* 37, 3472 (2003).

[68] D.M. Stevenson and P.J. Weimer, *Appl. Microbiol. Biotechnol.* 59, 721 (2002).

[69] D. Somayaji, Microbiological studies on biomethanation of mango peel with special reference to methanogens. Ph.D. thesis, University of Gulbrga, (1992).

[70] A. Pandey, C.R. Soccol, P. Nigam, and V.T. Soccol, *Bioresource Technol.* 74, 69 (2000).

[71] L.V. Reddy, O.V.S. Reddy and Y. Wee, *African J. Biotechnol.* 10, 4183 (2011).

[72] S. Couri, S.C. Terzi, G.A.S.S. Pinto, S.P. Freitas and A.C.A. Costa, *Process Biochem.* 36, 255 (2000).

[73] D.R. Kashyap, P.K. Vohra, S. Chopra and R. Tewari, *Bioresource Technol.* 77, 215 (2001).

[74] R.S. Jayani, S. Saxena, and R. Gupta, Microbial pectinolytic enzymes a review. *Process Biochem.,* 40, 2931 (2005).

[75] P.N. Padma, K. Anuradha, B. Nagaraju, V.S. Kumar and G. Reddy, *J. Bioprocess Biotechnol.* 2, 2 (2012).

[76] Y.S. Kumar, P. V. Kumar and O.V.S. Reddy, *Food Biotechnology,* 26, 107, (2012).

[77] S.S. Kumar and P. Palanivelu, *World J Microbiol. Biotechnol.* 15, 643 (1999).

[78] A. Thakur, R. Pahwa, S. Singh, and R. Gupta, *Enzyme Research*, 1 (2010).

[79] E.S. Martins, D. Silva, R. Da Silva, and E. Gomes, *Process Biochem.* 37, 949 (2002).

[80] M.E. Acuna-Arguelles, M. Gutierrez-Rojas, G. Viniegra-Gonzalez and E. Favela-Torres, *Appl. Microbiol. Biotechnol.* 43, 808 (1995).

[81] V.K. Joshi, M. Parmar, and N.S. Rana, *Food Technol. Biotechnol.* 44, 253 (2006).

[82] Y. S. Kumar, S. Varakumar and O.V.S. Reddy, *World J. Microbiol. Biotechnol.* 26, 1973 (2010).

[83] Devendra Kumar, M. Ashfaque, M., Muthukumar, S. Singh and N. Garg, *J. Env. Biol.* 33, 81 (2012).

[84] R.K. Sukumaran, R.R. Singhania and A. Pandey, *J Sci Industrial Res.* 64, 832 (2005).

[85] R. Carle, P. Keller, A. Schieber, C. Rentschler, T. Katzschner, D. Rauch, G. F. Fox and H.U. Endress, *Patent application*, WO 01/78859 A1. 2001.

[86] B. Bajpai, and S. Patil, *Enzyme Microbiol. Technol.* 20, 612 (1997).

[87] R. Paranthaman, R. Vidyalakshmi and K. Alagusundaram, *Academic J. Plant Sci.* 2, 124 (2009).

[88] C.G. Kumar and H. Takagi, *Biotechnol. Adv.* 17, 561 (1999).

[89] G. D. Haki and S. K..Rakshit, *Bioresource Technol.* 89, 17 (2003).

[90] P. Ellaiah, J. Premkumar, P.V. Kanthachari and K. Adinarayana, *Hindustan Antibiotics Bulletin*, *44*, 1 (2002).
[91] C. Sandhya, A. Sumantha, G. Szakacs, and A. Pandey, *Process Biochem.,* 40, 2689 (2005).
[92] A. Gessesse, *Bioresource Technol.* 62, 59 (1997).
[93] K. Aikat and B.C. Bhattacharyya, *Process Biochem.* 35, 907 (2000).
[94] P. Rathakrishnan and P. Nagarajan, *International J. Chem. Technol. Res.* 3, 1526 (2011).
[95] R.S. Prakasham, Ch.S. Rao and Sarma, P.N. *Bioresource Technol.* 97, 1449 (2006).
[96] T. Sathish, G.S. Lakshmi, Ch.S. Rao, P. Brahmaiah, and R.S. Prakasham, *Letters Appl. Microbiol.* 47, 256 (2008).
[97] M.K. Chimata, P. Sasidhar, and S. Challa, *African J. Biotechnol.* 9, 5162 (2010).
[98] H. Meruvu and M. Vangalapati, *Rasayan J Chemistry,* 4, 159 (2011).
[99] P. Iyer and R.S. Singhal, *Bioresource Technol.* 99, 4300 (2007).
[100] F.A. Schmid and J. Roberts, *Cancer. Chemother. Rep.* 58, 829. (1974).
[101] J. Zhao, A.L. Lopez, D. Erichsen, S. Herek, R.L. Cotter, N.P. Curthoys, and J. Zheng, *J. Neurochem.* 88, 169 (2004).
[102] A. Mulchandani and A.S. Bassi, *Biosens. Bioelectron.* 11, 271–280. (1996).
[103] A. Sabu, M. Chandrasekaran and A. Pandey, *Chemistry Today,* 18, 21 (2000).
[104] T. Nakadai and Nasuno, *J. Ferment.Bioeng.* 67, 158 (1989).
[105] T. Tachiki, T. Yamada, K. Mizuno, M. Ueda, J. Shiode and H. Fukami. *Biosci. Biotechnol. Biochem.* 62, 1279 (1998).
[106] F.J. Rispoli and V. Shah, *J. Industrial Microbiol Biotechnol.* 34, 349 (2007).
[107] P. Kashyap, A. Sabu, A. Pandey, G. Szakacs and C.R. Soccol, *Process Biochem.* 38, 307 (2002).
[108] G.N. Prabhu and M. Chandrasekaran, *Process Biochem.* 32, 285 (1997).
[109] M. Damien, J. Catherine, D. Patrice and B.Christopher, *Carbohydr. Poly.* 80, 747 (2010).
[110] R. Reeta, Anil Kumar., C.R. Soccol and A. Pandey, *Biochemical Engineering J.* 44, 13 (2009).
[111] D. Gangadharan, K. Madhavan Nampoothiri, S. Sivaramakrishnan, and A. Pandey, *Appl. Biochem. Biotechnol.* 158, 653 (2009).

In: Enzymes in Value-Addition of Wastes
Editors: Satinder Kaur Brar and Mausam Verma
ISBN: 978-1-62948-968-1

Chapter 5

PROTEASE: ENZYME-AIDED VALUE-ADDITION IN FOOD PROCESSING INDUSTRIES

***Ritu Raval*[*,1], *Keyur Raval*[2], *Satinder Kaur Brar*[3] and *Mausam Verma*[4]**

[1]Department of Biotechnology, Manipal Institute of Technology, Manipal University, Manipal, India
[2]Department of Chemical Engineering, National Institute of Technology Karnataka, PO Srinivasnagar, Mangalore, India
[3]INRS-ETE, Université du Québec, 490, Rue de la Couronne, Québec(QC), Canada
[4]CO_2 Solutions Inc., Québec, Canada

ABSTRACT

Environmental and economic apprehensions behoove a reduction in food processing wastes and processing of food residuals to new value added products, enzyme being one of the means of obtaining the same. In this chapter we intend to exemplify the initiatives engendered by different groups to confront the issue of waste disposal and searching a more environment friendly method of handling the bio wastes. This has been elaborated by the panoramic view of the waste generated worldwide. In addition different enzymes produced using the industrial waste has been discussed. In the subsequent section, the trailing towards protease has been made. In continuation, an initiative made by different food based industries in the employment of protease in degradation of the industrial wastes has been touched on upon in the first section of the chapter. In the second section, the research on the use of various organic wastes generated by the food industries for the protease production by solid state fermentation has been discussed. This would in turn help in value addition of the organic waste generated.

* E-mail: ritu.raval@manipal.edu.

1. Introduction

The food and agricultural-processing industry generates large volumes of wastes annually worldwide causing serious problems with their disposal. These problems are further aggravated in the countries which thrive largely on the agricultural practices. The scale of the problem is illustrated by looking at the total amounts of waste materials produced by different countries. Table 1 shows a list of waste quantities mentioned in the literature [1]. Currently, these agro-wastes are either allowed to decay naturally on the fields or are burnt. This action aggravates the environmental pollution burden the earth is already facing. Many steps have been taken by the academia and the industries to confront with management of the waste generated from these sectors. One such initiative is the utilization of the waste generated in the food industry as a raw material in the production of industrially relevant compounds utilizing solid-state fermentation (SSF) wherein the wastes which are rich in the sugar content are assimilated by the micro-organisms [Table 2]. In addition to catering to reducing environmental pollution, the utilization of these wastes would also reduce the production costs. Therefore, SSF is finding an increased application in the production of value added products from wastes [2]. Additionally, enzymes can be intermixed with the waste creating an environment-friendly method for waste management. The utilization of enzymes thus aids in tackling the reduction of food processing wastes in four main areas; i) the quantity of food waste decreases because they enable a better utilization of raw materials or commodities. ii) the enzymatic processing of waste streams gives a higher value by-products as exemplified by fish processing wastes, iii) they aid in improving the processing efficiency of raw materials, and iv) they assist in the clean-up of food wastes in by-product streams created during processing. Therefore, investigation and development of potential value-added processes for biological wastes is highly attractive. Microbes which are not only the major producers of enzymes but also of wide spectrum of substratum can be exploited in this respect. Of all the enzymes produced, protease occupies a pivotal role in the industrial application holding around 60% of the total sales. Therefore in the following chapter, a panoramic view of the world scenario with respect to the organic waste would be presented streaming it further to the usage of protease, an industrially important enzyme in the value addition of organic wastes.

Table 1. Quantities of organic wastes generated in different countries (Modified from [1])

Country	Quantity of waste generated [million tonnes/annum]	Type of waste
USA	251.3	Municipal waste[3]
China	150	municipal organic waste [4]
Brazil	95	municipal organic waste [5]
India	64.2	organic waste [6]
Canada	51.6	organic waste[7]
Japan	17	municipal organic waste [8]
Germany	3	Crude fiber residues from sugar production
	1.9	spent malt and hops [breweries]

Country	Quantity of waste generated [million tonnes/annum]	Type of waste
	1.8	grape pomace [viniculture]
	0.4	organic waste from potato, vegetable and fruit processing
	0.1	wet apple pomace [9]
Egypt	6.8	wheat straw residues
	1.6	Cotton stalks
	1.1	public garden waste
	0.7	vegetable residues
	0.7	date palm residues
	0.6	sesame straw[10]
France	9.7	organic waste [11]
Malaysia	4.1 2.5 1.4	oil palm shells empty fruit bunches Palm oil production palm mesocarp fibre [12]
Thailand	0.1	palm kernel shells
	0.1	palm press fibre,
	0.4	empty palm fruit bunches
	1	cassava pulp[12]
South Africa	1.4	organic waste [13]
Australia	0.4	pineapple peel [12]
Belgium	0.4	Organic waste [9]
Spain	0.3	olive pomace[9]
Italy	0.2	organic waste[9]
Greece	0.03	[dry weight] of citrus peels[9]
Portugal	0.01	tomato pomace[9]

Table 2. Production of industrially important enzymes from the agricultural wastes using SSF

Waste	Micro-organism	Product, Production scale, titer value	Reference
Sugarcane bagasse Wheat bran Sugarcane	*Aspergillus terreus,* *A. niger*	Xylanase, Lab scale with yield of 74500 IU/L	[14]
Wheat bran	Gibberella fujikuroi	Amylase, lab scale, yield of 50 IU/DMB	[15]
Wheat straw, barley straw, wood shaving	*Trametes versicolor*	Laccase, lab scale with yield of 2000 IU/L	[16]
Apple, orange and potato peelings	*Trametes hirsuta*	Laccase, lab scale with yield of 5000 IU/L	[17]
Wheat bran, rice husk	*Bacillus subtilis*	Amylase, 21760 IU/gm of rice husk	[18]
Spent brewing grains	*A. oryzae*	Amylase,Lab scale, 6410 IU/gm of substrate	[19]
Defatted soyabean cake	*Penicillium sp*	Protease, 55000 IU/L, Lab scale	[20]

Table 2. (Continued)

Waste	Micro-organism	Product, Production scale, titer value	Reference
Wheat bran	*Bacillus sp AS-1, A. sp AS-2*	Amylase, glucoamylase, 1,98,950 U/g fermented dry matter & 3426 U/g fermented dry matter	[21]
Wheat bran	*A. parasiticus*	Protease, 25000 IU/L at lab scale	[22]
Sugarcane Bagasse	*Thermoascus auriantacus*	Xylanase, Lab scale with activity of 1597 U/g	[23]
Coconut oil cake	*A.oryzae*	Amylase, 3388 U/gds	[24]
Barley bran, wheat straw	*T. versicolor, T. hirsuta*	Laccase, 181,000 IU/L	[25]
Mixture of wheat bran and lentil husk	*Bacillus* sp.	Alkaline protease, 429.024 IU/gm	[26]
Solid wheat bran and 10% purified chitin flakes	*P. aculeatum, P. chrysogenum, P. funiculosum, P. isariiforme*	Chitinase, 0.9 IU/GDS	[27]
Kiwi fruit wastes	*T. hirsute*	Laccase, 90,000 IU/L	[28]
Tamarind seed powder, palm kernel cake	*A. niger*	Tannase, 13.03 IU/gds	[29]
Wheat bran, rice husk, spent brewing grain coconut oil cake, palm kernel cake, sesame oil cake, jackfruit seed powder, olive oil cake	*A. oryzae, P. funiculosum*	Neutral protease, 8.7 IU/gds	[30]
Wheat bran, rice bran, corn bran,	*Bacillus sp.*	Amylase, 4,64,000 U/g dry bacterial bran	[31]
Sorghum straw	*Thermomyces anuginosus*	Xylanase, 48000 IU/gds	[32]
Rice bran	*Bacillus sp.*	Cellulase, 3644 IU/gds	[33]
Green gram husk, chick pea husk, red gram husk, black gram husk, wheat bran	*Bacillus sp.*	Alkaline protease, 10,000 IU/gds	[34]
Sawdust, wheat bran, rice bran, sugarcane, rice straw dust	*A. aculeatus*	Tannase, 174.32 IU/gds	[35]
Apple peel, Passion fruit peel, Lemon peel, Rice straw, Wheat bran, Corncob, Sugarcane bagasse	*A. niveus*	Pectin Lyase, 73000 IU/ml	[36]
Olive oil cakes	*Candida utilis*	Lipase and Protease, 25 IU/gm and 110 IU/gm respectively	[37]
Passion fruit waste	*Phanerochaete chyrsosporium*	Oxidative and Hydrolytic enzymes,	[38]

2. Proteases: An Introduction to the Enzyme Structure

Proteases, the enzyme catalyzing the cleavage of peptide bonds in other peptides are the single class of enzymes which occupy a pivotal position with respect to their applications in both physiological and commercial fields. The estimated value of the worldwide sales of proteases was $350 million in 2002 and was expected to increase at the rate of 7% annually [39]. The enzyme market grew at the rate of 12% per annum and it was US$ 5.1 billion in 2009 out of which industrial enzymes accounted for ca. US$ 2.7 billion. The major contribution to these sales is accounted by the hydrolytic enzymes with 75% of the shares. In this major contribution, the proteases represent one of the three largest groups of industrial enzymes and account for about 60% of the total worldwide sale of enzymes. According to the Nomenclature Committee of the International Union of Biochemistry and Molecular Biology, proteases are classified in subgroup 4 of group 3 (hydrolases) [40]. Currently, proteases are classified on the basis of three major criteria: (i) type of reaction catalyzed, (ii) chemical nature of the catalytic site, and (iii) evolutionary relationship with reference to structure. The catalytic site of proteases is flanked on one or both sides by specificity sub-sites, each able to accommodate the side chain of a single amino acid residue from the substrate. These sites are numbered from the catalytic site S1 through Sn toward the N terminus of the structure and Sl′ through Sn′ toward the C terminus. The residues which they accommodate from the substrate are numbered Pl through Pn and P1′ through Pn′, respectively (Figure 1, Figure 2).

Proteases execute a large variety of complex physiological functions. Their importance in conducting the essential metabolic and regulatory functions is evident from their occurrence in all forms of living organisms. Proteases play a critical role in many physiological and pathological processes such as protein catabolism, blood coagulation, cell growth and migration, tissue arrangement, morphogenesis in development, inflammation, tumor growth and metastasis, activation of zymogens, release of hormones and pharmacologically active peptides from precursor proteins, and transport of secretory proteins across membranes. In general, extracellular proteases catalyze the hydrolysis of large proteins to smaller molecules for subsequent absorption by the cell whereas intracellular proteases play a critical role in the regulation of metabolism. In contrast to the multitude of the roles contemplated for proteases, our knowledge about the mechanisms by which they perform these functions is very limited. In addition to the physiological activities, the protease also participate in the protein turnover, sporulation and conidial discharge, germination, enzyme modification, nutrition, regulation of gene expression [43- 46]

3. Value Addition of Wastes: Protease Aided Processes

Alkaline proteases hydrolyze proteins from diverse groups ranging from plants and animals to produce hydrolysates of well-defined peptide profiles and thus can add to the value of waste. An overview of its various roles in different industries has been detailed in Fig 3. In one such attempt, the commercial alkaline protease, Alcalase which possesses a broad specificity was used yielding a less bitter hydrolysate and a debittered enzymatic whey protein hydrolysate [47, 48].

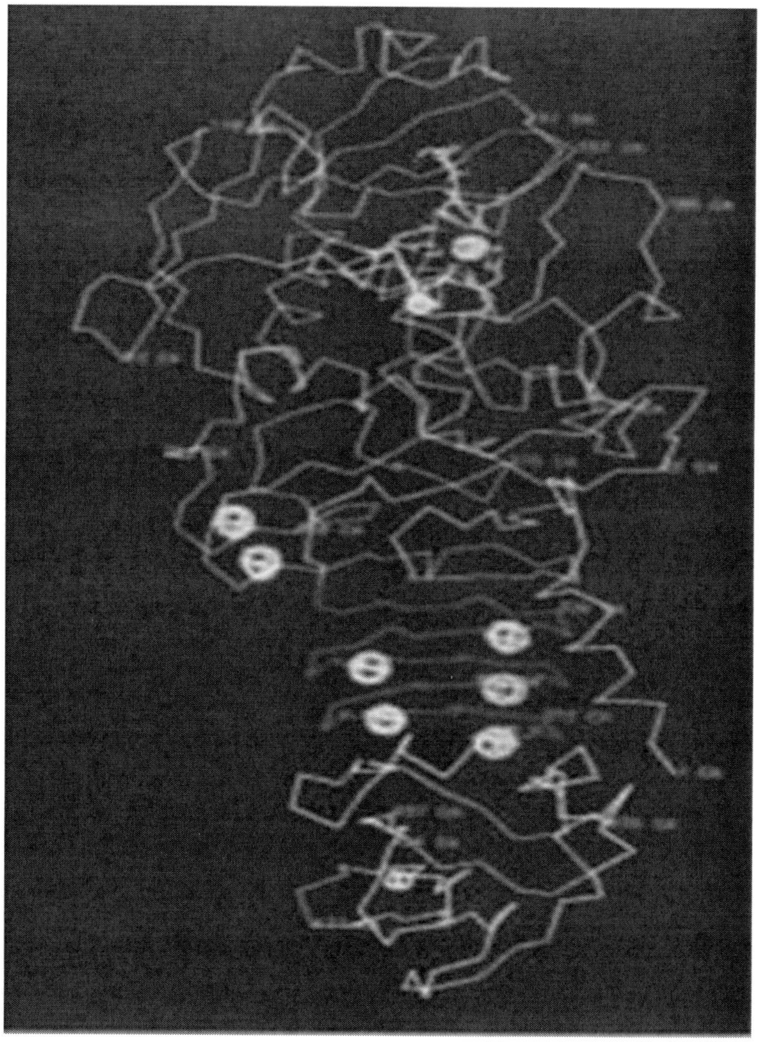

Figure 1. The crystallized structure of alkaline protease isolated from Pseudominas aeruginosa. The N-terminal helix is depictd in light blue, the proteolytic domain in green, C-terminal in yellow and the β-sandwich domain in red-violet. (Adapted from Mc. Kay etal., 1993, 41)

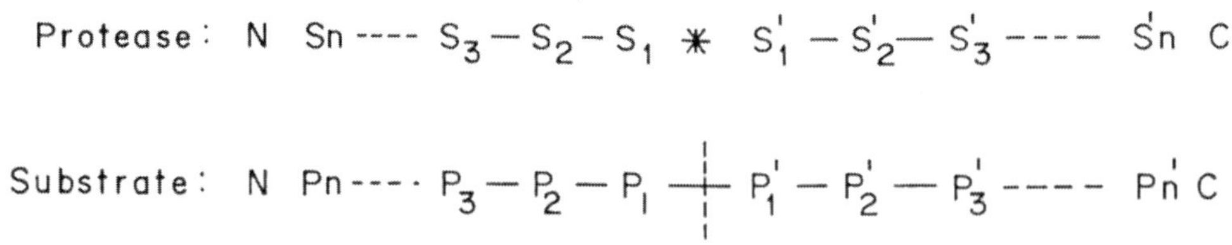

Figure 2. Active sites of proteases. The catalytic site of proteases is indicated by p and the scissile bond is indicated by Á ; S1 through Sn and S19 through Sn9 are the specificity subsites on the enzyme, while P1 through Pn and P19 through Pn9 are the residues on the substrate accommodated by the subsites on the enzyme (Adapted from Rao MB, 1998, 42).

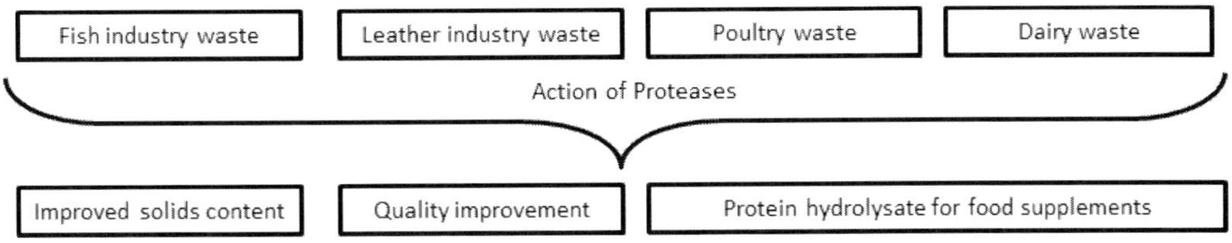

Figure 3. Schematic diagram elucidating the usage of protease in value addition of waste from various streams.

In another study an attempt using the alkaline protease from *B. amyloliquefaciens* resulted in the production of a methionine–rich protein hydrolysate from chick pea protein [49]. These hydrolysates specially the ones generated from casein, whey protein and soy protein find major application in hypoallergenic infant food formulations [50]. In addition the hydrolysates have also been used for the fortification of fruit juices or soft drinks and in maintaining protein–rich therapeutic diets [51, 52].

Furthermore, proteases play a prominent role in meat tenderization, especially of beef. An alkaline elastase [53, 54, 55] and thermophilic alkaline protease [56] have proved to be successful and promising meat tenderizing enzymes, as they possess the ability to hydrolyze connective tissue proteins as well as muscle fibre proteins.

In addition to its role in the meat tenderizing, the byproducts of the meat industry i.e. the skin hides also rely on the proteases. These dehaired hides are used by in the leather industry. The process of dehairing utilizes protease in place of the chemical method which is environmentally friendlier and commercially a more attractive option. Another application of the proteases is in the poultry industry which is a major section of the meat industry wherein the waste management of the feathers which makes up approximately 5% of the body weight of poultry is the major dilemma. The feathers can be utilized as a high protein source for food and feed, provided their rigid keratin structure is completely destroyed. A major initiative was made by Dalev [57] who used alkaline protease from *B. subtilis* for the management of waste feathers from poultry slaughter houses. In addition, the use of keratinolytic protease for food and feed industry waste, for degrading waste keratinous material from poultry refuse [58,59] and as depilatory agent to remove hair from the drains has also been reported [60].

The role of the protease is not only limited to the meat industry but has also been extended to the fish industry wherein a commercial protease preparation is used to prepare fish pellets. These recovered proteins are utilized as an additive (as a liquid concentrate or in dry formulation) for the nutritional value of the fish food. In addition to their role to produce nutritional additive, proteases have been added to liquid phase in fish meal production or stickwater, to reduce viscosity and increase the final solids content [61].

Further research to improve the protease action was contrived by a formulation constituting of the proteolytic enzymes from *B. subtilis*, *B. amyloliquefaciens* and *Streptomyces* sp. and a disulfide reducing agent (thioglycolate) to enhance hair degradation aiding in clearing pipes clogged with hair-containing deposits. This patented product sold as Genex is currently available in market [62].

4. Value Addition of Waste: Protease as a Commercial Product using SSF

Another growing field is the production of the protease in the agro or food wastes as a value addition to these wastes. Different waste sources and respective process conditions for protease production are described in the following figure.

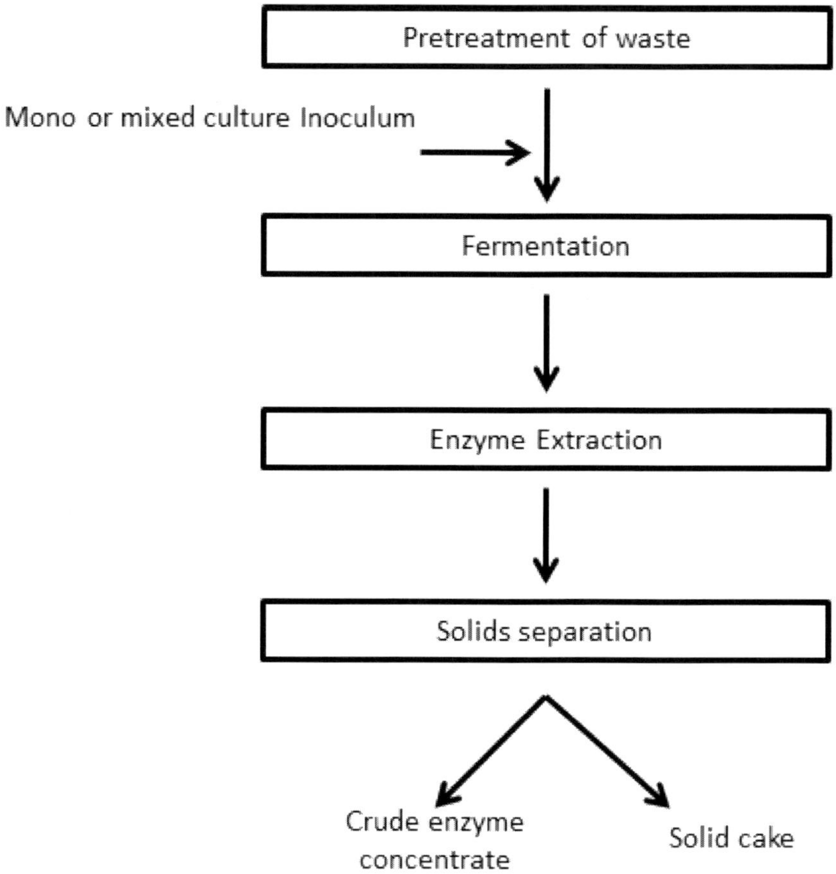

Figure 4. Schematic diagram of protease production process through fermentation route.

Lactose (From Dairy Waste)

In one of the propitious attempt, *Rhizopus oryzae* was grown on lactose as sole carbon source in whey besides fats and protein. This strain of *Rhizopus oryzae* isolated from the areal soil proved itself an efficient producer, greatly decreasing the BOD of whey in industrial effluent water [63].

Kitchen and Common Household Waste

The use of the protease has not only been limited to the diary waste but has also been extended to the kitchen waste. In one of the studies conducted on *Bacillus subtilis* DM-04 a thermophilic strain of under solid-state fermentation, different agro-industrial and kitchen waste materials, were screened for substrates/solid supports of alkaline protease production [64]. In the above study different combinations of the kitchen waste which was primarily potato peel with *I. cylindrica* grass was tested for protease production. The enzyme was not only thermostable but was also active in a range of pH. It also exhibited compatibility with most of the commercial laundry detergents, thus proving a good candidate for detergent formulation. Hence the value addition of the common kitchen waste can be improved for protease production.

Biodiesel Feedstock Residues

A major problem faced by after the oil extraction from *Jatropha curcas, is the large amount of by-products* generated from the seeds. An attempt was made to use the by-products for the protease production using SSF [65]. The study investigated the potential of deoiled *J. curcas* seed cake as substrate for protease productions by *Aspergillus oryzae*. While various parameters for SSF was individually optimized conventionally, five parameters were simultaneously examined based on the Taguchi method. The effect of three different levels of five factors, including moisture content of substrate, inoculums size, incubation temperature, type of porous substrate and incubation time were examined. By adjusting the conditions to these optimum levels, the protease production had improved by 4.6 times. In addition this work also proved to develop this agricultural waste for its industrial use.

Starch Waste from Food Crops

In one of the studies the possibility of utilizing cassava waste, one of the readily available substrates for protease production was exploited [66]. This study was conducted using response surface methodological approach for maximizing protease production. In another study conducted by Boominadhan etal., 2009 different agro wastes were studied for the impact on the protease activity with four different species of Bacillus viz *subtilis*, *megaterium*, *amyloliquefaciens* and *licheniformis* and with the agro waste ranging from soy cake, coconut cake, wheat bran, groundnut cake and brewery spent grain. It was found that maximum activity of the enzyme was detected with soy cake with *B. subtilis* helping yield the maximum activity while the coconut cake yielded in the lowest activity [67].

Role of Protease as Industrial Hygiene Agent

In addition to its role in the value addition of waste in the food industry, the proteases have also been tested for the cleaning in place (CIP) effect in the dairy industry by Prof.

Walsh [68] wherein 14 different strains of fungi were screened for their protease activity on milk protein for the removal of industrial-like milk fouling deposit from stainless steel. This was an attempt made to substitute the chemical based caustic based cleaning solutions at 70–80 C with the enzymatic treatment which in turn would result in potential environmental benefits with reduced energy consumption, decreased chemical usage and a reduced requirement for pH neutralization of the resultant waste prior to release.

Conclusion: Cost Comparison of Waste Treatment Versus Protease Production

Numerous methods such as dry feeding, wet feeding, landfill, incineration, anaerobic co-digestion, co-digestion by sewage sludge, composting etc. are employed for enormous amounts of waste generated from food processing industries and agricultural industries. Although variety of methods is available for disposal of organic waste, landfill option is predominant because of low costs associated with landfill option [4, 9, 12, 69]. Various reports suggest landfill option as the most inefficient in terms of benefits and environmental impact [69-71]. The greenhouse gas emissions are the largest in landfill option apart from other environmental issues related to leaching and air pollution. As per the table 1 tremendous amounts of organic waste generation in developed and developing countries necessitates alternative methodologies for use of such waste to reduce greenhouse gases and adverse environmental impact. Production of protease enzymes using organic waste seems very promising option as it will not just impart value-addition to the waste but also reduce adverse environmental impacts. Raw material costs are about 30 to 40% of the total operational costs of protease production plant [39, 70-71]. Use of waste as raw material for protease production will increase benefit margins, reduce emissions of harmful gases to the environment because of conversion of carbon and nitrogen in useful products and thus will have tremendous impact on environment.

References

[1] Laufenberg, G., Kunz, B., Nystroem, M., Transformation of vegetable waste into value added products: (A) the upgrading concept; (B) practical implementations. *Bioresour.Technol.* 2003, *87*, 167–198.

[2] Raghavarao, K. S., Ranganathan, M.S., Karanth, T.V., Some engineering aspects of solid-state fermentation. *Biochem. Eng. J.* 2003, *13*, 127–135.

[3] Kollikkathara,N., Feng, H., Stern, E., A purview of waste management evolution: Special emphasis on USA. *Waste Management*. 2009, *29*, 974-985.

[4] Chen, X.,Geng, Y.,Fujita,T., An overview of municipal solid waste management in China, *Waste Management,* 2010 *30*, 716–724.

[5] Lino, F.A.M, Ismail, K.A.R, Cosso, I.L, Evaluation of the potential of recycling for the reduction of energy and CO2 emissions in Brazil. Sustain. *Cities Soc.,* 2013.

[6] Waste management in India, www.ebtc.eu

[7] National Inventory Report – Greenhouse Gas Sources and Sinks in Canada 1990-2005, Greenhouse Gas Division, Environment Canada, Environment Canada. 2007a.

[8] Morisaki, I. *Pattern of Food Losses in Households: A Case Study in Oita-Prefecture,* Japan. Oita, Japan: Ritsumeikan Asia Pacific University, 2011.

[9] Mazzanti, M., Zoboli, R.,. Waste generation, waste disposal and policy effectiveness: Evidence on decoupling from the European Union. Resources, *Conservation and Recycling* 2008 *52*, 1221–1234.

[10] Shaban D., Hussein A., Omaima, M., The Utilization of Agricultural Waste as One of the Environmental Issues in Egypt, 2010, *J Appl Sci Res, 6,*1116-1124.

[11] *Waste figures in France,* ADEME, 2009

[12] Dijkgraaf, E., Herman R., Vollebergh, J,. Burn or bury? A social cost comparison of final waste disposal methods. 2004, *Ecol Eco, 50*, 233-247.

[13] Nahmana, A., de Langea, W., Oelofseb, S., Godfrey, L., The costs of household food waste in South Africa, 2012, Waste Management, *32*, 2147–2153.

[14] Gawande, P. V., Kamat, M. Y., Production of *Aspergillus* xylanase by lignocellulosic waste fermentation and its application. *J. Appl. Microbiol.* 1999, *87*, 511–519.

[15] Mulimani, V. H., Ramalingam, P. G. N., Amylase production by solid state fermentation: A new practical approach to biotechnology courses. *Biochem. Educ.* 2000, *28*, 161–163.

[16] Rodríguez Couto, S., Gundín, M., Lorenzo, M., Sanromán,A., Screening of supports and inducers for laccase production by *Trametes versicolor* in semi-solid-state conditions. *Process Biochem.* 2002, *38*, 249–255.

[17] Rosales, E., Rodríguez Couto, S., Sanromán, A., New uses of food wastes: Application to laccase production by *Trameteshirsuta. Biotechnol. Lett.* 2002, *24*, 701–704.

[18] Baysal, Z., Uyar, F., Aytekin, C., Solid state fermentation for production of α-amylase by a thermotolerant *Bacillus subtilis* from hot-spring water. *Process Biochem.* 2003, *38*, 1665–1668.

[19] Francis, F., Sabu, A., Nampoothiri, K. M., Ramachandran, S.*et al.*, Use of response surface methodology for optimizing process parameters for the production of α-amylase by *Aspergillus oryzae*. *Biochem. Eng. J.* 2003, *15*, 107–115.

[20] Germano, S., Pandey, A., Osaku, C. A., Rocha, S.N., Soccol, C. R., Characterization and stability of proteases from *Penicillium* sp. produced by solid-state fermentation. *Enzyme Microb.Tech.* 2003, *32*, 246–251.

[21] Soni, S. K., Kaur, A., Gupta, J. K., A solid state fermentation based bacterial α-amylase and fungal glucoamylase system and its suitability for the hydrolysis of wheat starch. *Process Biochem.* 2003, *39*, 185–192.

[22] Tunga, R., Shrivastava, B., Banerjee, R., Purification and characterization of a protease from solid state cultures of *Aspergillus parasiticus*. *Process Biochem.* 2003, *38,* 1553–1558.

[23] Milagres, A. M. F., Santos, E., Piovan, T., Roberto, I. C., Production of xylanase by *Thermoascus aurantiacus* from sugar cane bagasse in an aerated growth fermentor. *Process Biochem.* 2004, *39*, 1387–1391.

[24] Ramachandran, S., Patel, A. K., Nampoothiri, K. M., Francis, F. *et al.*, Coconut oil cake-a potential raw material for the production of α-amylase. *Bioresour. Technol.* 2004, *93*, 169–174.

[25] Rodríguez Couto, S., Rosales, E., Gundín, M., Sanromán, M. Á., Exploitation of a waste from the brewig industry for laccase production by two *Trametes* species. *J. Food Eng.* 2004, *64*, 423–428.

[26] Uyar, F., Baysal, Z., Production and optimization of process parameters for alkaline protease production by a newly isolated *Bacillus* sp. under solid state fermentation. *Process Biochem.* 2004, *39*, 1893–1898.

[27] Binod, P., Sandhya, C., Pusztahelyi,T., Szakacs,G., Production and purification of extracellular chitinases from *Penicillium* spp. under solid state fermentation. *Enzyme Microb. Tech.* 2005, *36*, 880–887.

[28] Rosales, E., Rodríguez Couto, S., Sanromán, M.A., Reutilisation of food processing wastes for production of relevant metabolites: Application to laccase production by *Trametes hirsuta*. *J. Food Eng.* 2005, *66*, 419–423.

[29] Sabu, A., Pandey, A., Daud, M. J., Szakacs, G.,Tamarind seed powder and palm kernel cake: Two novel agro residues for the production of tannase under solid-state fermentation by *Aspergillus niger* ATCC 16620. *Bioresour. Technol.* 2005, *96*, 1223–1228.

[30] Sandhya, C., Sumantha,A., Szakacs,G., Pandey, A., Comparative evaluation of neutral protease production by *Aspergillus oryzae* in submerged and solid-state fermentation. *Process Biochem.* 2005, *40*, 2689–2694.

[31] Sodhi, H. K., Sharma, K., Gupta, J. K., Soni, S. K., Production of a thermostable α-amylase from *Bacillus* sp. PS-7 by solid state fermentation and its synergistic use in the hydrolysis of malt starch for alcohol production. *Process Biochem.* 2005, *40*, 525–534.

[32] Sonia, K. G., Chadha, B. S., Saini, H. S., Sorghum straw for xylanase hyper-production by *Thermomyces lanuginosus* (D2W3) under solid state fermentation. *Bioresour. Technol.* 2005, *96*, 1561–1569.

[33] Virupakshi, S., Gireesh Babu, K., Gaikwad, S. R., Naik,G. R., Production of a xylanolytic enzyme by a thermoalkaliphilic *Bacillus* sp. JB-99 in solid state fermentation. *Process Biochem.* 2005, *40*, 431–435.

[34] Prakasham, R. S., Rao,C. S., Sarma,P.N., Green gram husk— An inexpensive substrate for alkaline protease production by *Bacillus* sp. in solid-state fermentation. *Bioresour. Technol.* 2006, *97*, 1449–1454.

[35] Banerjee, D., Mondal, K. C., Pati, B. R., Tannase production by *Aspergillus aculeatus* DBF9 through solid-state fermentation. *Acta Microbiol. Immunol. Hung.* 2007, *54*, 159–166.

[36] A. Maller1, T. M. da Silva2, A. R. L. Damásio1, V. R. A. Reis2, J. A. Jorge2, M. L. T. M. Polizeli, Production of Pectin Lyase by Aspergillus niveus under Submerged and Solid State Fermentations Using Agro-Industrial Residues as Carbon Sources. *Int. Res. J. Microbiol.* 2012, *3*, 29-35.

[37] Moftah,O.A.S., Grbavčić,S., Žuža, M.,Luković, N., Bezbradica,D., Knežević-Jugović, Z., Adding Value to the Oil Cake as a Waste from Oil Processing Industry: Production of Lipase and Protease by *Candida utilis* in Solid State Fermentation. *Appl. Biochem. Biotechnol.* 2012, *166*, 348-364.

[38] Zilly, A., Bazanella,G.C. Cristiane, Helm,V., Aparecida Vaz Araújo,C., Giatti, C., de Souza, M., Bracht, A., Peralta, R., Solid-State Bioconversion of Passion Fruit Waste by White-Rot Fungi for Production of Oxidative and Hydrolytic Enzymes. *Food Bioprocess Technol.*,2012,*5*, 1573-1580.

[39] Kirk, O., T.V. Borchert and C.C. Fuglsang, Industrial enzyme applications. 2002.Curr. Opin. Biotechnol., *13*: 345-351.

[40] International Union of Biochemistry. Enzyme nomenclature. Orlando, Fla: Academic Press, Inc.; 1992.

[41] Baumann, U., Wu, S.,Flaherty, K.M., McKay,D.B., Three-dimensional structure of the alkaline protease of Pseudomonas aeruginosa: a two-domain protein with a calcium binding parallel beta roll motif., *EMBO J.*,1993, 12,: 3357–3364.

[42] Rao, M.B., Tanksale, A.M., Ghatge, M.S., Deshpande, V.V., Microbiol Molecular and biotechnological aspects of microbial proteases, *Mol. Biol. Rev.*, 1998, *62*, 597–635.

[43] Chung, C. H., and A. L. Goldberg, The product of the *lon* (*capR*) gene in *Escherichia coli* is the ATP dependent protease, protease La. Proc. Natl. Acad. Sci. USA,1981, *78*,4931–4935.

[44] Smeekens, S. P., Processing of protein precursors by a novel family of subtilisin-related mammalian endoproteases. *Bio/Technology,*1993,*11*,182–186.

[45] Kalisz, M. H., Microbial proteinases. *Adv. Biochem. Eng. Biotechnol.* 1988,*36*,17–55.

[46] Van Melderen, L., M. Thi, P. Leechi, S. Gottesman, M. Couturier, and M. L. Maurizi. ATP-dependent degradation of Ccd A by Lon protease, *J. Biol. Chem.* 1996, *271*,27730–27738.

[47] Adler-Nissen, J., *Enzymatic Hydrolysis of Food Proteins. New* York:Elsevier Applied Science Publisher,1986.

[48] Nakamura, T., Syukunobe, Y., Sakarai, T., Idota, T., Enzymatic production of hypoallegenic peptides from casein, *Milchwissenschaft,*1993, *48*, 11-14.

[49] George, S., Sivasankar, B., Jayaraman, K., Vijayalakshmi, M.A.,Production and separation of the methionine rich fraction from chick pea protein hydrolysate generated by proteases of *Bacillus amyloliquefaciens*, *Process Biochem.* 1997, *32*,401-404.

[50] AAPCN(American Academy of Pediatrics Committee on Nutrition) Hypoallergenic infant formulas. *Pediatrics,* 1989,*83*, 1068-1069.

[51] Adamson, N.J., Reynolds, E.C., Characterization of casein phosphopeptides prepared using Alcalase: determination of enzyme activity. *Enzyme Microb. Technol.*,1986, *18*, 202-207.

[52] Parrado, J., Bautista, J. Machado, A., Production of soluble enzymatic protein hydrolysate from industrially defatted non dehulled sunflower meal, *J. Agric. Food Chem.*, 1991, *39*, 447-450.

[53] Takagi, H., Kondou, M., Hisatsuka, T., Nakamori, S., Tsai, Y.C., Yamasaki, M., Effects of alkaline elastase from an alkalophilic Bacillus strain on the tenderization of beef meat, *J. Agric. Food Chem.*, 1992, *40*, 2364-2368.

[54]Takagi, H., Tsai, Y.C., Nakamori, S., Yamasaki, M., Improved production and recovery of alkaline elastase from alkalophilic *Bacillus* strain by a change of medium composition. *Biosci. Biotechnol. Biochem.*,1995, *59*, 1591-1592.

[55] Takagi, H., Protein engineering on subtilisin. *Int. J. Biochem.* 1993, *25*, 307-312.

[56] Wilson, S.A., Young, O.,A., Coolbear, T., Daniel, R.,M., The use of proteases from extreme thermophiles for meat tenderization, *Meat Sci.,* 1992, *32*, 93-103.

[57] Dalev, P.G., Utilization of waste feathers from poultry slaughter for production of a protein concentrate. *Bioresour. Technol.*, 1994, *48*:265–267.

[58] Ichida, J.M., L. Krizova, C.A. Lefevre, H.M. Kerner, D.L. Elwell and E.H. Jv Burtt: Bacterial inoculum enhances keratin degradation and biofilm formation in poultry compost. *J. Microbiol. Methods*,2001, *47*, 199-208.

[59] S. Agrahari, S., Wadhwa, N., Isolation and characterization of feather degrading enzymes from *Bacillus megaterium* SN1 isolated from Ghazipur poultry waste site. *Appl. Biochem. Microbiol.*, 2012, *48,*175-181.

[60] Takami, H., Nakamura, S., Aono, R., Horikoshi, K., Degradation of human hair by a thermostable alkaline protease from alkalophilic *Bacillus* sp. No. AH-101. *Biosci. Biotechnol. Biochem.* (1992d), *56,*1667–1669.

[61] Guerard, F., Guimas, L., Binet, A., Production of tuna waste hydrolysates by a commercial neutral protease preparation, *J. Mol. Catal B : Enzy.*,2002, *19-20*, 489-498.

[62] Jacobson, J.W., Glick, J.L., Madello, K.L., Composition for cleaning drains clogged with deposits containing hairs. US Patent 4,540,506, 1985.

[63] Gul, A., Baig, S., Naz, M., Nadeem, M., Efficient utilization of diary industry waste for hyper-production and charcaterization of a novel cysteine protease. *Pak. J. Zool.* 2012,*44*,713-721.

[64] Mukherjee,A., K., Adhikari, H., Rai S.K., Production of alkaline protease by a thermophilic *Bacillus subtilis* under solid-state fermentation (SSF) condition using *Imperata cylindrica* grass and potato peel as low-cost medium: Characterization and application of enzyme in detergent formulation, *Biochem. Engg. J.*, 2008, *39*,353–361.

[65] Thanapimmetha, A., Luadsongkram, A., Titapiwatanakun, B., Boosaree, Srinophakun, P.,Value added waste of *Jatropha curcas* residue: Optimization of protease production in solid state fermentation by Taguchi DOE methodology, *Industrial Crops Prod.*, 2012, *37*, 1–5.

[66] Boominadhan, U., Rajakumar, R., Sivakumaar, P.K.V., Joe, M.M., Optimization of protease enzyme production using *Bacillus* sp. isolated from different wastes, *Bot. Res. Int.*,2009, *2*, 83-87.

[67] Rathakrishnan, P., Nagarajan, Kannan, R., Optimization of process parameters using a statistical approach for protease production by *Bacillus subtilis* using cassava waste *Int. J. Chem. Tech. Res.,* 2012, *4*, 749-760.

[68] Boyce, A., Walsh, G., Identification of fungal proteases potentially suitable for environmentally friendly cleaning-in-place in the dairy industry, *Chemosphere*,2012, *88*, 211–218.

[69] Mi-Hyung Kim,Yul-Eum Song, Han-Byul Song, Jung-Wk Kim,Sun-Jin Hwang. Evaluation of food waste disposal options by LCC analysis from the perspective of global warming: Jungnang case, South Korea. *Waste Management,* 2011, *31*, 2112–2120.

[70] Krishna Prasad. *Downstream process technology: a new horizon in biotechnology.* PHI learning.

[71] Haddar, A., Fakhfakh-Zouari, N., Hmidet, N., Frikha, F., Nasri, M., Sellami Kamoun, A., Low-cost fermentation medium for alkaline protease production by Bacillus mojavensis A21 using hulled grain of wheat and sardinella peptone. *J Biosc. Bioeng.* 2010, *110*, 288–294.

In: Enzymes in Value-Addition of Wastes
Editors: Satinder Kaur Brar and Mausam Verma
ISBN: 978-1-62948-968-1

Chapter 6

DIFFERENT ENZYMES AND THEIR PRODUCTION

Naceur M'hamdi[1,*], Cyrine Darej[1], Jamel Jebali[2], Rachid Bouraoui[3], Sabrine Metahni[3], Ibtihel Frouja[2], Dhillon Gurpreet Singh[4], Imène Jarboui[4] and Satinder Kaur Brar[5]

[1]Institut National Agronomique de Tunis. Tunisie
[2]Institut Supérieur de Biotechnologie de Monastir, Université de Monastir. Tunisie
[3]Ecole Supérieure d'Agriculture de Mateur, Bizerte, Tunisie
[4]Institut Supérieur Agronomique de Chott-Meriem, Université de Sousse, Tunisie
[5]Institut National de la Recherche Scientifique, Centre Eau, Terre and Environnement, Québec. Canada

ABSTRACT

A large amount of wastes are generated every year from the industrial processing of agricultural raw materials. Most of these wastes are used as animal feed or burned as alternatives for their disposal. Agricultural wastes have huge potential for the production of enzymes. They are renewable form of resources generated round the year all over the world. These materials are accumulated in enormous amounts alone and serve as source of environmental pollution. Due to its abundant availability, it can serve as an ideal substrate for microbial processes for the production of enzymes. Agricultural processing inevitably goes along with the production of large amounts of agro-residues, which may represent a major waste disposal problem. In recent years, there has been an increasing trend towards efficient utilization and value-addition of agro-industrial residues such as wheat bran, coffee pulp, sugarcane bagasse, apple pomace and others. The production of enzymes by bioprocesses is a good alternative to add value to agroindustry residues. The bioconversion of the agro waste material into fuel has received considerable interest over the recent years. This lignocellulosic material is present in abundance and available free of cost. Enzymatic hydrolysis of cellulosic biomass is considered as the most efficient and least polluting methods for generating glucose from lignocellulosics. The agricultural wastes represent a significant potential for the development of biorefineries in different

[*] Corresponding author: Naceur M'hamdi. E-mail: naceur_mhamdi@yahoo.fr.

sectors, such as cereals. Recovery of phytochemicals as well as the energetic valorization of the plant matrices needs the demolition of the cell walls of plants.

Use of agroindustrial residues as substrates in fermentation processes provides an alternative avenue and value-addition. The concept of utilizing excess biomass or waste from agricultural and agro-industrial residues to produce enzymes is not necessarily new. Recently, fermentation of biomass has gained considerable attention because of the forthcoming scarcity of fossil fuels, and as it is necessary to increase the world food and feed supplies, especially those high in protein.

Keywords: *Aspergillus niger*, biomass, solid state fermentation, enzyme activity, cellulose, xylanase

INTRODUCTION

The worldwide food, agricultural and forestry industries produce large volumes of wastes, which cause a serious disposal problem (Rodríguez-Couto, 2008). Waste output and byproducts are inherent to all productive sectors. With the improvement of ecological awareness by the end of the 20th century, it became clear that human kind's major challenge for the coming decades was to balance the production of goods and services with economic growth, social equality and environmental sustainability (Galembeck et al., 2009). Agro-industrial residues are generally considered best substrates for enzyme production (Ellaiah et al., 2002; Ghosh and Ray, 2010). Nowadays, enzyme production is a growing field of biotechnology especially related to industrially important enzymes. With the advancement in biological sciences, man has realized the importance of bioconversion of industrial wastes into enzymes through the knowledge of sweeping role of micro-organisms. Meanwhile, biotechnology is finding its applications in major industrial domains, including health care, crop production and agriculture, non food (industrial) use of crops and other products and environmental uses. In view of rapidly growing economic demand, it is imperative to exploit all indigenous resources at their maximum to attain the goal. The production of enzymes from agro-industrial wastes can reduce to some extent the cost of production (Durrani et al., 2002). The agro industrial wastes are frequently used as substrates in fermentation as an inexpensive source. In food biotechnology, food industrial wastes are used for the production of enzymes. Fungi are recognized for their capacity to produce a variety of extracellular proteins, organic acids and other metabolites, and for their capacities to adapt to severe environmental constraints (Cochrane, 1958). For example, *A. niger* is the prototypical fungus for the production of citric acid (Lal, 1980 and Grewal and Kalra, 1995), and proteins (Wongwicharn et al., 1999 and Xu et al., 2000). Cellulose, the major constituent of all plant materials, forms about half to one-third of plant tissues and is constantly replenished by photosynthesis. One of the significant applications of agro-industrial wastes, such asapple pomace is biotechnological production of enzymes, such as cellulases and chitosan. Gao et al. (2010) proved that lignocellulosic biomass is known to be an excellent carbon source for microbial enzyme production. Various agricultural substrates, byproducts and microbial cultures have been successfully used in solid state fermentation (SSF) and submerged state fermentation (SMF) for cellulose and chitosan production (Duenas et al., 1995). In recent years, there has been an increasing trend towards more efficient utilization of agro-industrial residues, including apple pomace. Several processes and products have been reported that utilize apple pomace as a raw material.

Apple pomace is the processing waste generated after apple juice manufacturing. This solid residue consists of a complex mixture of peel, core, seed, calyx, stem, and soft tissue. This residual material is a poor animal feed supplement because of its extremely low protein content and high amount of sugar. The application of agro industrial by-products in bioprocesses offers a wide range of alternative substrates, thus mitigate pollution problems related to their disposal. Attempts have been made to use apple pomace to produce enzymes (Bhalla and Joshi, 1993). Biotechnological innovations bring significant and successful efforts to convert lignocellulosic residues into enzymes for biotechnological and industrial applications (Leite et al., 2008; Tengerdy and Szakacs, 2003). This chapter intends to bring enzymes together with the knowledge obtained from different biotechnological applications of agro-industrial wastes to enzymes.

Agro-Industrial Wastes

Agricultural, industrial, forestry and urban solid residues are abundant sources of lignocellulose available worldwide that were treated as wastes in many countries (Dashtban et al., 2009). Agro-industrial wastes and byproducts are renewable form of resources generated round the year all over the world. These materials are still treated as wastes, disposed in the environment, many times without an adequate treatment, raising environmental damage. They are also used to generate thermal energy (Dawson and Boopathy, 2007; Kim and Dale, 2004). Agricultural production and the agro-food industry furnish large volumes of solid wastes, residues and by-products, produced either in the primary agro-forestry sector (crop-based) or by secondary processing industries (processing-based), the major part being lignocellulosic biomass. Agro-industrial wastes are generated during the industrial processing of agricultural or animal products.

Those derived from agricultural activities include materials, such as straw, stem, stalk, leaves, husk, shell, peel, lint, seed/stones, pulp or stubble from fruits (Cavalitto et al., 1996; Martins et al., 2000; Silva et al., 2002), legumes or cereals (rice, wheat, corn, sorghum, barley…), bagasse generated from sugarcane or sweet sorghum milling, spent coffee grounds, brewer's spent grains, and many others.

Apple Pomace

Apple pomace is a waste resource produced in large quantities and its deposit has caused serious environmental problems. Though, it is an interesting raw material and has attracted considerable attention as a potential sugar, dietary fiber, pectin, and phenolic source. Sato et al. (2010) reported that these products can be used for many purposes (pharmaceutical, cosmetic and food industries).

Dietary fiber is conventionally classified in two categories according to their water solubility, insoluble dietary fiber (IDF) such as cellulose and lignin; and soluble dietary fiber (SDF) such as pectin, gums, and mucilage (Esposito et al., 2005). Furthermore, apple pomace is the main byproduct produced in the apple fruit processing industry. It is a rich source of carbohydrate, pectin, crude fiber, and minerals (Mirzaei-Aghsaghali and Maheri-Sis, 2008).

Schieber et al. (2003) reported that the dietary fiber content in apple pomace ranged between 33 to 35%, representing an important natural dietary fiber source. And its use for further product recovery is an appropriate measure for economic reasons as well as from an environmental point of view. This waste is destined, in many countries, for animal feeding or it is discharged without any other treatment causing environmental pollution, however it can be processed to generate value added products. Furthermore, IDF from apple pomace can be converted into SDF by enzymatic treatment. This is accompanied with an increase in the free phenol concentration, water-soluble antioxidant activity, and bioavailability of the phenol compounds as well as improving the sensory properties of pomace (Schieber et al., 2003; Napolitano et al., 2006). A successful recovery process which can easily be integrated into existing production processes must be established and production of alternative products should be explored with the purpose of having an economically feasible apple pomace processing (Shalani and Gupta, 2010). Industrial apple pomace is composed of press residue from cider apples, wines, brandies, distilled or spirits and vinegars (Smock and Neubert, 1950) as well as components from the residual epidermis and endocarp obtained in the semi-industrial processes of freezing, canning, dehydration, and other processing (Virk and Sogi, 2004). Drying apple pomace seems to be the most economically viable approach for stabilization as it significantly reduces the volume and leads to lower transportation costs. The drying yield at 60ºC is around 50 g/kg in 10 hours, or 5% from raw material. The appearance of dried pomace is dependent on the drying temperature. Wosiacki and Sataque (1987) confirmed that from 50 to 60°C, the enzymatic browning reactions were stimulated, while from 90 to 100°C, Maillard reactions occur, with products appearing darker than those obtained in range from 70 to 80°C. However, if the criterion to stop the process is the time when the pomace temperature begins to raise, such temperature will never be higher than 52°C and the final product appearance tend to be homogeneous. The instability of apple pomace is related to its physicochemical composition and to the presence of enzymes activated after plant tissue disintegration (Kennedy et al., 1999; Endreb, 2000). Apple pomace contains average moisture of 70.5% and 10.8-15 % reducing sugars (table 1). The fiber content varies from 4.7 to 51.1%, and includes cellulose (7.2 g/kg) and lignin (23.5g/kg) (Dhillon et al., 2011).

Orange Peel and Pulp

Citrus fruits constitute an important group of fruit crops produced all over the world. The production of citrus pulp pellets requires a large capital investment by the processor with little if any return on investment. Due to high capital costs, many small citrus juice processors are unable to install feed mills to produce citrus pulp pellets, leaving them with tons of peel waste for disposal that has little if any value (Mark et al., 2005). Citrus fruits are utilized mostly for table purposes, but a significant portion is processed into various products, such as squashes, cordials, single strength juices, juice concentrates, marmalades, pickles etc. Dried citrus peel is rich in carbohydrates, proteins and pectin; the fat content, however, is low (Vibha and, Neelam, 2011). Citrus fruits comprise an important group of fruit crops manufactured worldwide. In the fruit processing industry large amounts of waste materials are produced, in the form of peel, pulp and seeds (Dhillon et al., 2004).

The waste material presents significant disposal difficulties, and when not used in any way it causes odor and soil pollution (Dhillon et al., 2004, Ma et al., 1993). Since the 1980s the worldwide production of citrus has increased drastically. Estimations show that in 2010 the orange production reached 66.4 million metric tons, which was an increase of 14% compared with that of 1997-1999. Almost half of the 30.1 million metric tons of the produced orange will be manufactured to yield juice, essential oils and other by-products (Mamma et al., 2008). While dried citrus peels are rich in cellulose, hemicelluloses, proteins (5.25%) and pectin, the fat content is however, low (Dhillon et al., 2004; Ma et al., 1993). In the citrus processing industry, citrus peels are the major solid by-product and comprise around 50% of the fresh fruit weight. The citrus waste can be used as raw material for pectin extraction or in pelletized form for animal feeding. However, the citrus waste has to be dried first, and none of these processes has been found to be very profitable (Mamma et al., 2008). A disadvantage is that orange peels have a very low nutritional content which reduce its value as livestock feed (Dhillon et al., 2004).

CEREALS STRAW

Straws, as well as other fibrous by-products, have been utilized as animal feed in the whole world for centuries. However, the constraints resulting from the low nitrogen (N) and the high fiber contents, as well as the increased importance of concentrate feeds has led to a continuous decrease in its utilization. When offered to livestock both dry matter intake and palatability are low. However, these by-products are most often the only available feed in tropical and subtropical areas where smallholder mixed crop livestock systems prevail. More recently, Shiere et al. (2004) have showed that straws and stoves are increasingly important, e.g. for animal feed, thatching, soil improvement, mushroom production and industrial use. According to Duarte et al. (2010), the fiber content of cereals straw includes cellulose (30%), hemicelluloses (50%) and lignin (15%). Wheat straw is one of the most abundant crop residues in European countries with a production of 170 million tones and seems to be the cheapest and the most useful raw material for ethanol production (Fang et al., 2002). It is comprised mainly of cellulose (40%), hemicelluloses (26%) and about 20% of lignin (Lequart et al., 1999). Hydrolysis of cellulose and hemicelluloses is hampered by the overlapping of polymer embedded microfibrils.

SUGARCANE

Sugarcane is one of the important cash crops in India and plays pivotal role in both agricultural and industrial economy of the country. The production of sugar from sugarcane yields vast amounts of biomass, especially in the form of molasses, vinasse, and bagasse. Value-added products from bagasse are of interest. Conversion of lignocelluloses residues, such as bagasse into furfural is an old established technology that has been employed at many plants. Sugarcane bagasse has a great biotechnological potential, which is partially due to its abundance, since it is one of the largest agro-industrial by-products, being produced in over 200 countries.

Every year, more than 400 million tons of sugarcane bagasse is produced around the world (UNSD, 2005). Peel, pulp and membranes from oranges and other related citrus fruits are highly susceptible to hydrolysis by mixtures of cellulolytic and pectinolytic enzymes (Grohmann and Baldwin, 1992). Sugarcane bagasse (SCB) is a fibrous residue of cane stalks left over after the crushing and extraction of the juice from the sugarcane is one of the largest cellulosic agro-industrial by-products. Sugarcane is composed essentially of juice and fibers. The juice is an aqueous solution of sugars (meanly sucrose and small fractions of fructose and glucose mono-saccharaides) and other organic and inorganic substances (minerals and impurities). Fiber is defined as all insoluble material in cane. The composition of sugarcane stalks depends on a large number of factors, including the age of the cane, growing conditions, disease, while the composition of sugarcane delivered to the factory depends on stalk composition, cane variety, age, amount of tops and leaves and other exogenous matter carried in harvesting operation (Rein, 2007). Composition-wise, the average ash contents of 2.4% are attractive when employing bioconversion methods using microbial intervention (Pandey et al., 2000). Cellulose content is typically 39-42% and lignin 20-25%. Hemi-cellulose has been commonly observed in the 25% to 27% range (Gámez et al., 2006; Bustos et al., 2003; Yu and Stahl, 2008), although higher values of 36% have also been reported (Sasaki et al., 2003). Among other biotechnological applications, it is particularly interesting to note that bagasse has been used for the production of industrially important enzymes, such as cellulases and xylanases, the same enzymes that are responsible for an effective degradation of bagasse and other lignocellulosic materials (Parameswaran, 2009; Adsul et al., 2004).

MICROORGANISMS

A large number of microorganisms including bacteria, yeast and fungi produce different groups of enzymes. Selection of a particular strain however, remains a tedious task, especially when commercially competent enzymes, yield are to be achieved. Microorganisms, such as bacteria and fungi are widely exploited for the industrial production of numerous enzymes. A large variety of microorganisms are capable of degrading plant cell wall materials (Carle Urioste et al., 1997). Filamentous fungi (moulds) can grow on a wide range of substrates and efficiently degrade biopolymers and are thus, an attractive resource for new enzymes. Micro-organisms have been closely associated with transforming or cycling organic matter in nature for as long as such material has existed. Microorganisms have long played a major role in the production of food (dairy, fish and meat products) and alcoholic beverages. In addition, several products of microbial fermentation are also incorporated into food as additives and supplements (antioxidants, flavours, colourants, preservatives, sweeteners...). Recently, a large number of microorganisms, isolated from different materials, have been screened for their ability to degrade polysaccharide present in vegetable biomass producing cellulases. There is a great interest in the development and use of natural food and additives derived from microorganisms, since they are more desirable than the synthetic ones produced by chemical processes. A diverse spectrum of lignocellulolytic microorganisms, mainly fungi

(Baldrian and Gabriel, 2003; Falcón et al., 1995) and bacteria (Zimmermann, 1990; Vicuňa, 1988) have been isolated and identified over the years and this list still continues to grow.

Aspergillus Niger

Aspergillus niger is a member of the genus *Aspergillus* which includes a set of fungi generally considered asexual and commonly found as a saprophyte growing on dead leaves, stored grains, compost piles, and other decaying materials. *A. niger* is being used in food industry for the production of different enzymes like α-amylase, cellulases, lactase, amyloglucosidase, invertase, pectinases and proteases that advocates its safety regarding food applications (Immanuel et al., 2007; Maciel et al., 2008). The genus, A*spergillus is* economically, ecologically and medically important.

It is ubiquitous in nature with over 185 species. The fungus owes its name from the Latin word aspergillum, to the resemblance of its conidiophores to a device used to sprinkle holy water. *Aspergilli* colonies are downy to powdery in texture. The surface color varies depending on the species.

Microscopically, the asexual fruiting structure of *Aspergillus* species includes a long stipe (conidiophores), a rounded head (vesicle), and flask shape structures (phialides) from which spores (conidia) are formed. The conidia (2-5 μm in diameter) may be spherical to elongate and form chains which may radiate or form themselves into compact columns. *A. niger* has been used for the production of cellulase and chitosan.

Trichoderma Reesei

Trichoderma reesei has been studied for several years and it has long been acknowledged that when growing the fungus on e.g. cellulose, addition of a complex organic nitrogen source increases the enzyme production. It was reported by many authors that *Trichoderma reesei* is an efficient producer of cellulase enzymes. And strains of *T. reesei* have been studied in great detail for more than two decades for their ability to produce the cellulase enzyme complex (Kadam, 1996; Philippidis, 1994).

Most literature reports describe experiments that have achieved a significant enzyme titer or a high volumetric productivity. Filamentous fungi *Trichoderma reesei* are considered to be one of the most efficient hyper producers of cellulase that is used in industry. *Trichoderma reesei* is a biotechnically important filamentous fungus used commercially in enzyme production. *T. reesei* is also one of the best known cellulolytic organisms, producing readily and in large quantities a complete set of extracellular cellulases for the degradation of crystalline cellulose.

In addition to *T. reesei*, a wide variety of other bacteria and fungi also produce cellulolytic enzymes (Miettinen-Oinonen, 2004). The genus *Trichoderma* comprises a group of filamentous ascomycetes that are widely used in industrial applications because of their ability to produce extracellular lignocellulose-degrading hydrolases in large amounts. Enzymes secreted by *Trichoderma* have received widespread industrial interest, leading to commercial applications in the textile industry (Cavaco-Paulo and Gűbitz, 2003; Nierstrasz and Warmoeskerken, 2003), the food and feed industries (Galante et al., 1998) and the pulp and paper industry (Buchert et al., 1998).

Trichoderma reesei has the capacity to secrete enzymes in high yields and this property can be exploited when using *T. reesei* as an industrial host for homologous and heterologous enzyme production (Penttilä et al., 2004).

Trichoderma Viride

Trichoderma viride is one of many species of mold. Most molds are innocuous saprophytes, living off dead organic materials, although a number of them are plant pathogens. The mycelium of *Trichoderma* can produce a wide variety of enzymes, including cellulases (degrading cellulose) and chitinases (degrading chitin). Due to its cellulases, *Trichoderma* can grow directly on wood, which is primarily composed of cellulose, a polymer of glucose. Due to their chitinases, *Trichoderma* can be a parasite of other fungi. *Trichoderma viride* are popular as it contains high activities of both exo-glucanase and endo-glucanase but low levels of β-glucosidase (Latifian et al., 2007).

Bacillus Subtilis

Bacillus species continue to be dominant bacterial workhorses in microbial fermentations. *Bacilus subtilis* is one of the most widely used bacteria for the production of specific chemicals and industrial enzymes and also a major source of amylase and protease enzymes. *Bacillus subtilis* produces a variety of extracellular enzymes including proteases, amylases and lipolytic enzymes of great importance in industrial processes, such as pharmaceutical, leather, laundry, food and waste processing industries (Schallmey et al., 2004; Machius et al., 1995). *Bacillus subtilis* (natto) is the key microbial participant in the ongoing production of the soya-based traditional natto fermentation, and some *Bacillus* species are on the Food and Drug Administration's GRAS (generally regarded as safe) list. *Bacillus* species have been major workhorse industrial microorganisms with roles in applied microbiology, which date back more than a thousand years, since the production of natto by solid-state fermentation of soybeans using *Bacillus subtilis* (natto) was first practiced in Japan (Hara and Ueda, 1982). Natto or the participating culture, *B. subtilis* (natto), is reported to have a number of beneficial health effects (Ueda, 1989). Natto contains antibiotic properties that are reported to be due to a heat unstable bacterial lytic activity and to dipicolinic acid. The consumption of natto during lactation was reported to increase vitamin K in breast milk. *Bacillus subtilis* (natto) cultures have been reported to eliminate pathogenic bacteria from the intestinal canal, and fractions from extracts of natto exhibited anticancer activity and reduced blood pressure in specific laboratory trials. *Bacillus subtilis* and related *Bacillus* strains continue to be the dominant enzyme-producing microorganisms in applied and industrial microbiology (Gupta et al., 2002). These organisms are an important source of industrial extracellular enzymes, including proteases and amylases, and the biochemistry and physiology of these strains have been investigated in detail. Because of the proven capability of these species to produce and secrete gram quantities per litre of these and other enzymes, they have been considered as potential candidate hosts for the production of heterologous proteins. While *B. subtilis* (natto) produces many enzymes, including amylases and cellulases, the most important enzymes in the production of natto are proteases; two proteases having a pH optima of 8.5 and 10.3-10.8 have been characterized (Yoshimoto et al. 1971). The proteases are responsible for the main flavor, through hydrolysis of soybean protein. *Bacillus* species are great candidates as hosts for the production of heterologous proteins due to their high capacity to secrete proteins into the extracellular medium.

Wang and Doi (1989) described the various genetic signals, required for effective transcription and translation of foreign genes in *B. subtilis*, and the requirements to overcome barriers to gene expression.

TRANSFORMATION

The development of transformation techniques for *Aspergillus* and other filamentous fungi lagged behind those for *E. coli* and *S. cerevisiae*, complicated by their multicellular morphology, thick chitinous cell walls, and lack of plasmids. The first reports of gene cloning (Kinghorn and Hawkins, 1982) and transformation in *Aspergillus* (Ballance and Buxton, 1983) were in 1982 and 1983, respectively. Several laboratory methods for *Aspergillus* transformation are currently in use. The most popular technique, adapted from that originally developed for yeast, is the chemical treatment of nucleated proto-plasts produced by enzymatic digestion of the fungal cell wall. Refinements of this method have increased transformation efficiencies several orders of magnitude since its conception from 10 transformants/μg DNA to several hundred or thousand for some systems (Dawe et al., 2000). Other methods include electroporation (Sanchez and Aguirre, 1996) and a biolistic technique (Barcellos et al., 1998). The plant transformation workhorse *Agrobacterium tumefaciens* has been used to transform *Aspergillus* and other fungi (Gouka et al., 1999), including economically important species, such as the common edible button mushroom, *Agaricus bisporus*. As critics of GMO foods have pointed out, transformation with bacterial vectors may carry some risk associated with integration of extraneous vector DNA flanking the gene of interest, e.g., antibiotic selection markers. Thus, the possibility to integrate only desired genes into the genomes of drug producing or food species could pave the way for rapid development of this method. By the mid-1990s, a range of *Aspergillus* transformation and expression vectors had been developed, most based on nutritional selection markers known from classical genetic studies (Hynes, 1996).

FERMENTATION

Fermentation is the oldest platform for biological conversion and continues to constitute the most fundamental and mature area of biotechnology. For thousands of years, fermentation was important for preserving and processing food and beverages. The fermentation is a core of biotechnology and comprises mainly two types: solid-state (SSF) and submerged fermentation (SMF) (Pandey, 2003). Traditionally, SSF technologies have been widely practiced in the manufacture of fermented foods, such as soy sauce, sake etc. The Western World use SSF for the manufacture of antibiotics and enzymes but now it has been replaced with SMF, which is a superior technology in terms of automation, containment and large volume fermentation. Fermentation processes can be carried out in two ways; surface or submerged production. SSF involves the growth of microorganisms on moist solid substrates in the absence of free water. This low moisture content makes SSF different from SMF. Unlike the situation in SMF, there is no systematic study guiding the design and operation of large scale SSF with proper controls.

Solid State Fermentation (SSF)

The most common process for the production of enzymes involves SMF using *Aspergillus niger*, growing on media containing glucose or sucrose (Leangon et al., 2000 and Kumar et al., 2003). Romero-Gomez et al. (2000) confirmed that the use of SSF has shown some potential as an alternative for the production of this organic acid. Robinson et al. (2001) approved that SSF involves the growth of microorganisms on moist solid substrates in the absence or near-absence of free flowing water.

Likewise, Nandakumar et al. (1994) demonstrated that the solid substrate acts as a source of carbon, nitrogen, minerals and a growth surface which absorbs the water necessary for microbial growth. Solid substrate provides anchor points for the growth and propagation of microorganisms. As microorganisms on a solid substrate are growing under conditions similar to their natural habitat, they may be able to produce enzymes, metabolites, proteins and spores more efficiently than in submerged fermentation (Goes and Sheppard, 1999). In recent years, SSF has regained interest based on its technological and economic advantages over conventional SMF.

The absence of a liquid phase and low substrate humidity level allow facilitated aeration through the pore spaces between substrate particles, reduced risk of bacterial contamination, simplified media and utilization of agro-industrial sugar rich wastes or by-product (Gutierrez-Correa et al., 1999). Various agro-industrial residues, such as apple pomace, coffee husks, wheat straw, pineapple waste, cassava baggasse, banana peel, sugar beet and kiwi fruit peel have been used as solid substrates.

Submerged Fermentation (SMF)

Submerged production began in the 1930's and is the main method used today. Submerged fermentation consists of the cultivation of microorganisms in liquid nutrient broth. This involves growing carefully selected microorganisms (bacteria and fungi) in closed vessels containing a rich broth of nutrients and a high concentration of oxygen. As the microorganisms break down the nutrients, they release the desired enzymes into solution. Due to the development of large-scale fermentation technologies, the production of microbial enzymes accounts for a significant proportion of the biotechnology industry total output. Fermentation takes place in large vessels (fermenter) with volumes of up to 1 m^3. And the fermentation media sterilizes nutrients based on renewable raw materials like maize, sugars and soya.

Most industrial enzymes are secreted by microorganisms into the fermentation medium in order to break down the carbon and nitrogen sources. Parameters, such as temperature, pH, O_2 consumption and CO_2 formation are measured and controlled to optimize the fermentation process. At the outset, in harvesting enzymes from the fermentation medium one must remove insoluble products (microbial cells).

This is normally done by centrifugation. As most industrial enzymes are secreted by cells into the external environment, they remain in the fermented broth after the biomass has been removed. The biomass can be recycled as a fertilizer, but first it must be treated with lime to inactivate the microorganisms and stabilize it during storage.

The enzymes in the remaining broth are then concentrated by evaporation, membrane filtration or crystallization depending on their intended application. If pure enzyme preparations are required, they are usually isolated by gel or ion exchange chromatography.

Potential Applications of Agro-Industrial Wastes for the Production of Enzymes

Agro-industrial wastes are mainly composed of complex polysaccharides that might serve as nutrients for microbial growth and production of enzymes. There are different products that can be produced from this agro-industry for agricultural uses, such as plague attack, drought, nutritional deficit, phytotoxicity from agrochemicals, growth regulation, fungus inhibitors, substitute of chemical pesticides, biofertilizers, etc. Rodríguez-Couto (2008) has recently published an article on the exploitation of large scale biological wastes for the production of as many as 19 types of enzymes.

The agricultural raw materials have been used for the bioconversion of their appropriate components for the production of enzymes (Koutinas et al., 2007). The surplus agricultural materials have been used for the production of enzymes (Koutinas et al., 2007; Venus and Richter 2006). The production of enzymes has evolved rapidly and nowadays, enzymes are the most important products obtained for human needs through microbial sources. Enzymes are substances present in the cells of living organisms in minute amounts and are capable of speeding up chemical reactions (associated with life processes), without themselves being altered after the reaction. They accelerate the velocity of biochemical reactions (Sajitha et al., 2011). Enzymes have application in a variety of areas including food biotechnology, environment, animal feed, pharmaceutical, textile, paper and others technical and chemical industries (Solange et al., 2012).

Cellulase

Cellulase is an enzyme complex which breaks down cellulose to β-glucose (Figure 1). It is produced mainly by symbiotic bacteria in the ruminating chambers of herbivores. A side from ruminants, most animals (including humans) does not produce cellulase, and are therefore unable to use most of the energy contained in plant material. Cellulase refers to family of enzymes which act in concert hydrolyzes cellulose.

Cellulose is widely distributed throughout the biosphere and is most manifest in fungal and microbial organisms. Three general types of enzymes make up the cellulose complex. Endocellulase breaks internal bonds to disrupt the crystalline structure of cellulose and expose individual cellulose polysaccharides chains. Exocellulase cleaves 2-4 units from the ends of the exposed chains. Cellobiase or beta-glucosidase hydrolyses the endocellulase product into individual monosaccharide. It digests fiber and it helps remedy digestive problems such as mal absorption. It is a very important enzyme because the human body cannot produce it on its own.

Also it helps in the breakdown of plant wall (cellulose, and increase the overall efficiency of binding excess cholesterol and toxins in the intestines for removal.

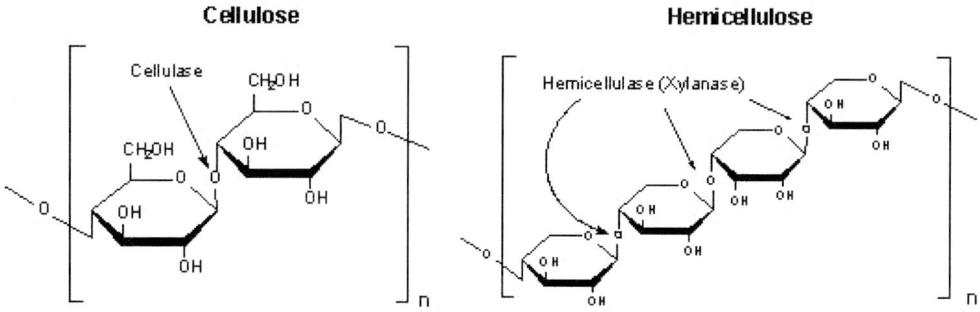

Drawn from Aldrich, 2007.

Figure 1. Representative structure of cellulose and hemicellulose chains.

It may be beneficial for food and environmental allergies, drug withdrawal, colon cleaning and pain syndromes, yeast infections, gas, bloating, acute food allergies, facial pain or paralysis. It is used in the fermentation of biomass into biofuels, although this process is relatively experimental at present. Otherwise, cellulase is used in animal healthcare as a feed supplement for better milk yield, in poultry and cattle industry. Most cellulase studied has similar pH optima, solubility and amino acid composition. However, it should be remembered that cellulase preparations generally contain other enzymatic activates besides cellulase, and these may affect the properties of the preparations.

XYLANASE

Xylanases are the group of enzymes that are involved in the hydrolysis of xylans and arabinoxylan polymers. These enzymes include endo-1, 4-β-xylanase, β-xylosidase, a-arabinofuranosidase and acetylxylan esterase (Biely, 1993). Xylanases hydrolyze 1, 4-β-D-xylosidic linkages in xylan to produce xylo-oligosaccharide (Figure 2).

Xylan is the major hemicellulose component and approximately accounts 20-35% of plant cell wall dry weights (Jiang et al., 2010). The heterogeneity of xylan led to a diversity of xylan-degrading enzymes (Dwivedi et al., 2009). Xylanases have considerable interest because of their applications in food processing, bleaching of pulp during paper industry, bio-conversion of biomass wastes to fermentable sugars and clarification of fruit juices (Romanowska et al., 2006). They also have application in improvement nutrient digestibility in animal diets and in the production of 3xylooligosaccharides and xylose (Shah and Madamwar, 2005).

During the last few decades, significant interest has developed in xylan and its hydrolytic enzymatic complex for use in bread production, as a supplement in animal feed, and in the preparation of drinks, textiles, bleaching of cellulose pulp, ethanol, and xylitol (Javier et al., 2007; Coughlan and Hazlewood, 1993). Wong et al. (1988) classified microbial xylanases into two groups based on their physicochemical properties such as molecular mass and isoelectric point, rather than on their different catalytic properties. While one group consists of high molecular mass enzymes with low pI values the other of low molecular mass enzymes with high pI values, but exceptions are there.

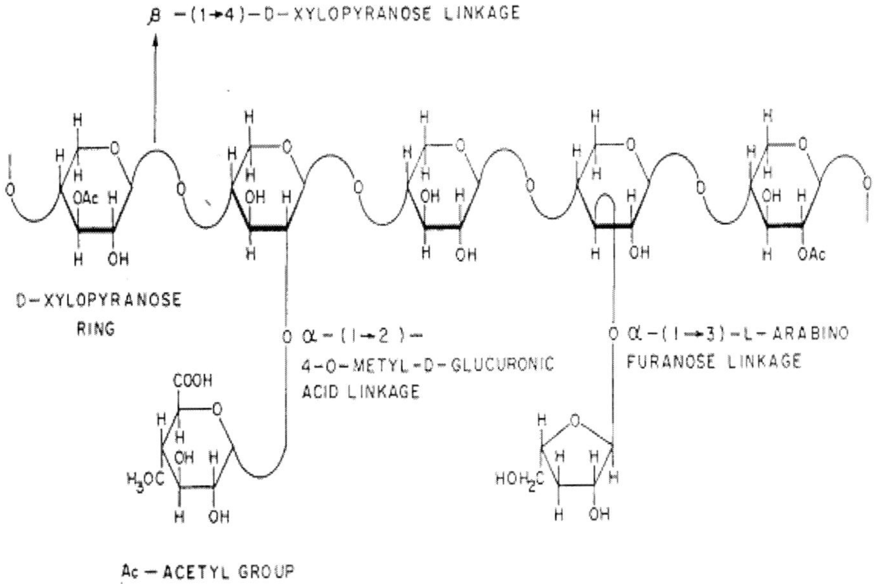

Belly, 1985 in Bastawde, 1992.

Figure 2. Xylan structure with side chains attached.

The above observation was later found to be in tune with the classification of glycanases based on hydrophobic cluster analysis and sequence similarities (Sapag et al., 2002).

Potential application of xylanases in biotechnology include biobleaching of wood pulp, treating animal feed to increase digestibility, processing food to increase clarification and converting lignocellulosic substances to feedstock and fuels (Subramaniyan and Prema, 2002).

CHITOSAN

Chitin, a linear polysaccharide composed of (1/4)-a linked N-acetyl-b-D-glucosamine unit is an extremely insoluble material and so far has been found very little large-scale industrial use (Figure 3).

Chitin can be converted to chitosan by deacetylation. Chitosan is soluble in acid solutions, and has a wide range of use being employed in the cosmetics, pharmaceuticals, agricultural and food industries. It can be used for clarification of juices (Imeri and Knorr, 1998) and production of biodegradable packaging films (Mayer et al., 1989). Its use as a component of toothpaste, hand and body creams, shampoo, cosmetics and toiletries, as well as pharmaceuticals, for lowering of serum cholesterol has been documented.

Furthermore, its application in enzyme (Synowiecki and Al-Khateeb, 2003) and cell immobilization, as a drug carrier, a material for production of contact lenses or eye bandages, as well as seed coats etc, have all been reviewed (Muzzarelli, 1989).

In addition, it can be applied to deliver drugs to the target (Onishi et al., 1997) and used as flocculating and chelating agents (Wan et al., 2004).

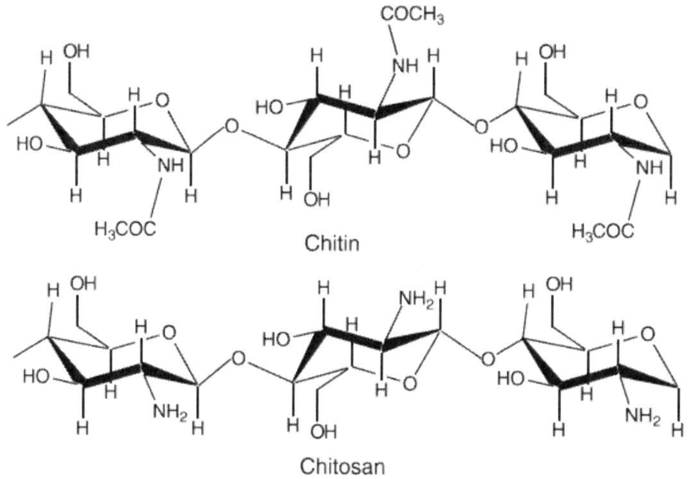

Piza et al., 1999.

Figure 3. Chemical structure of a monomer of chitin or chitosan.

LACCASES

Laccase (*p*-diphenol:oxygen oxidoreductase, EC 1.10.3.2) multicopper protein belonging to the family of the blue oxidase enzymes (Lee et al., 2003). It is a polyphenol oxidase, which belongs to the family of blue multicopper oxidases. Laccase was first discovered in the sap of the Japanese lacquer tree *Rhus vernicifera* in 1883 (Reinhammar, I984). It is a polyphenol oxidase, which belongs to the family of blue multicopper oxidases.

It is a polyphenol oxidase, which belongs to the family of blue multicopper oxidases. This enzyme generally contains four copper ions, grouped in three groups: T1, formed by one ion and is responsible for substrate oxidation and for electron transfer; T2, also formed by one ion and, together with the T3 group, which contains two ions, constitute the copper trinuclear center, involved in oxygen reduction and water release (Torres et al., 2003) (Figure 4).

Laccase is the most common ligninmodifying enzyme produced by the white-rot fungi belonging to the family Polyporaceae *sensu lato* (Pelaez et al., 1995). Laccase is classified as a phenol oxidase and catalyzes the oxidation of various aromatic and inorganic compounds (phenols in particular) and at the same time reduces oxygen to water. Laccases are the mostly studied enzymes produced from lignocellulosic wastes (Couto and Sanromána, 2006; Mishra and Kumar, 2007; Alcántara et al., 2007).

Laccases are the most important components of the lignolytic complex of wood-destroying white rot fungi responsible for decomposition (but in some cases also synthesis) of lignin, which is one of the most widely distributed natural polymers.

Laccases are able to depolymerize lignin (Kawai et al., 1999) and delignify wood pulps (Bourbonnais et al., 1997) when they are combined with various low molecular weight electron transfer agents. Hence, laccase is potentially applicable for pulp bleaching (Srebotnik and Hammel, 2000) and detoxification of phenolic pollutants (Collins et al., 1996; Johannes et al., 1996; Hublik and Schinner, 2000).

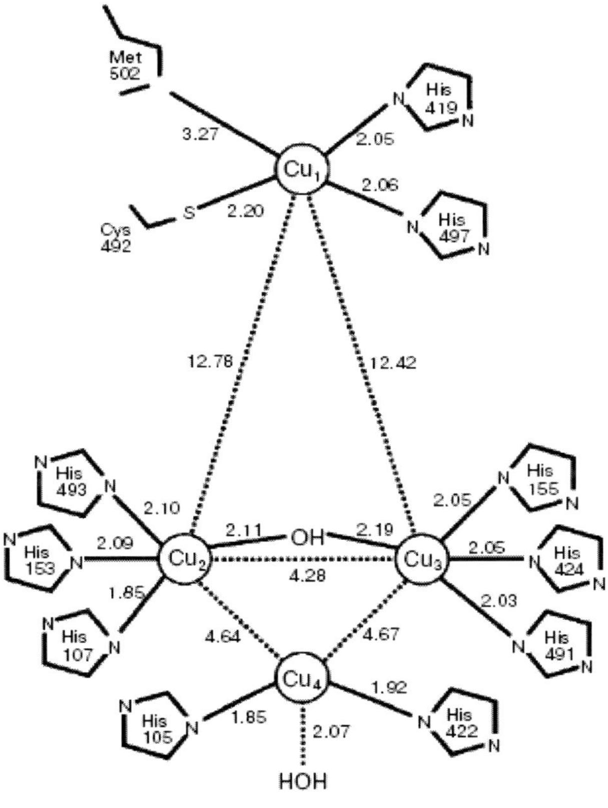

Enguita et al., 2003.

Figure 4. Scheme of T1 (Cu1) and T2/T3 (Cu4/Cu2-Cu3) copper sites of laccase CotA from *Bacillus subtilis*, with indicated distances between the most important atoms.

AMYLASES

Alpha-Amylase (EC 3.2.1.1) also named as 4-α-D-glucan glucanohydrolase enzymes are important enzymes employed in starch processing industries for hydrolysis of polysaccharides such as starch into simple sugar constituents (Hashim et al., 2005; Akpan et al., 1999; Nigam and Singh, 1995).

Amylases are a group of enzymes produced by several micro-organisms including fungi (Fadel, 2000; Wang et al., 2001). They are commercially used for processing starch to produce sugar syrups containing glucose, maltose and higher oligosaccharides (Hagihara et al., 2001). These enzymes account for about 30% of the world's enzyme production (Van der Maarel et al., 2002).

Pandey et al. (2000) and Gupta et al. (2003) reported that alpha-Amylas, has found its application in a range of industries including food, brewing, distilling industry, textile, paper pharmaceutical and bioconversion of solid waste etc. And large range of applications is the triggering factor for the industrialization of alpha amylase production.

CONCLUSION

The advent of biotechnology, specifically fungal biotechnology, with its inexpensive mode of application, has been used as tool for the effective conversion of agro-based wastes into certain enzymes (Ahuja et al., 2004). Increasing use of enzymes to replace traditional chemical transformation processes is obsessed by an aspiration for better production economics, new product functionalities, improved safety, and an increasing desire to reduce the environmental pollution.

Oftenly, chemical transformations produce non-specific reactions that may result in poor product yields and byproducts that are difficult and costly to dispose of agro-industrial wastes are generated in large amounts every year. Finding alternatives for the reuse of these wastes is an objective that has been strongly taken into account by countries around the world, considering environmental and economical aspects.

The processes involving organic solvents, acidity or alkalinity at high temperatures and pressures need expensive, specially designed equipment and control systems. To meet the increasing demand for enzymes, most new enzymes are produced from fungal or bacterial kingdom grown in large-scale fermenters using agroindustrial wastes (Cherry et al., 2001). Bioconversion of agro-industrial wastes for production of enzymes has received considerable interest in recent years.

Majority of industrial enzymes are produced by large-scale submerged fermentation. Agro-industrial wastes can eventually be used as substrate and act as good sources of carbon and nitrogen. This involves growing selected microorganism in closed vessels in which all the conditions critical for growth are carefully controlled. Selected microorganisms are either bacteria (*Bacillus* species) or fungal (*Aspergillus* or *Trichoderma* species) that have been carefully chosen and optimized.

REFERENCES

Adsul, M., Ghule, J., Singh, R., Shaikh, H., Bastawde, K., Gokhale, D., Varma, A. (2004). Polysaccharides from bagasse: applications in cellulase and xylanase production. *Carbohydrate Polymers*, 57: 67-72.

Ahuja, S. K., G. M. Ferreira and A. R. Moreira. 2004. Utilization of enzymes for environmental applications. *Crit. Rev. Biotechnol.* 24(2-3): 125-54.

Akpan, I., Bankjole, M. O., Adesermowo, A. M., and Lantunde-Data, 1999. Production of α-amylase by *Aspergillus niger* in a cheap solid medium using rice bran and agricultural material. *Trop. Sci.*; 39; 77-79.

Alcántara, T., Góme, J., Pazos, M., Sanromán, Á., 2007. Enhanced production of laccase in Coriolopsis rigida grown on barley bran in flask or expanded-bed bioreactor. *World J. Microbiol. Biotechnol.* 23(8): 189-1194.

Aldrich, S. 2007. Enzymes for Carbohydrate Analysis and Digestion. *Dairy Science*, 82: 996-1003.

Baldrian, T., Gabriel, J., 2003. Lignocellulose degradation by *Pleurotus ostreatus* in the presence of cadmium. *FEMS Microbiol. Lett.* 220: 235-240.

Ballance and Buxton, 1983. Transformation of *Aspergillus nidulans* by the orotidine-5'-phosphate decarboxylase gene of Neurospora crassa. *Biochem. Biophys. Res. Commun.*, 112 (1) (1983), pp. 284–289.

Barcellos, F. G., Fungaro, M. H. P., Furlanto, M. C., Lejeune, B., Pizzirani- Kleiner, A. A., and Azevedo, J. L., 1998. Genetic analysis of Aspergillus nidulans unstable transformants obtained by the biolistic process. *Can. J. Microbiol.* 44(12): 1137–1141.

Bastawde, K. B., 1992. Xylan structure,microbial xylanases, and their mode of action. *World Journal of Microbiology and Biotechnology* 8, 353-368.

Biely, P. 1993. Biochemical aspects of the microbial hemicellulases; In: *Hemicelluloses and Hemicellulase*, Coughlan, M. and Hazlewood, G. (eds.), pp. 29-51, Portland Press, London, UK.

Bourbonnais, R., Paice, M. G., Freiermuth, B., Bodie, E., Borneman, S., 1997. Reactivities of various mediators and laccases with kraft pulp and lignin model compounds. *Applied Environmental Microbiology* 63, 4627-4632.

Buchert, J., Oksanen, T., Pere, J., Siika-aho, M., Suurnäkki, A., and Viikari, L. 1998. Applications of Trichoderma enzymes in the pulp and paper industry. In: Harman, G. and Kubicek, C. (eds.) *Trichoderma* and *Gliocladium, Enzymes, biological control and commercial applications,* Vol. 2:343-364.

Bustos, G., Ramírez, J., Garrote, G., Vázquez, M. (2003), Modeling of the Hydrolysis of Sugarcane Bagasse with Hydrochloric Acid. *Applied Biochemistry and Biotechnology*, 104:51-68.

Carle-Urioste, J. C., Escobar-Vera, J., El-Gogary, S., Henrique-Silva, F., Torigoi, E., Crivellaro, O., Herrera-Estrella, A., El-Dorry, H., 1997. Cellulase induction in Trichoderma reesei by cellulose requires its own basal expression. *J. Boil. Chem.* 272: 10169-10174.

Cavaco-Paulo, A. and Bitz, G. 2003. Catalysis and processing. In: Cavaco-Paulo, A. and G¸bitz, G. (eds.) *Textile processing with enzymes.* Wood head Publishing Ltd., England. pp. 86-119.

Cavalitto, S. F., J. A. Arcas and R. A. Hours, 1996. Pectinase production profile of Aspergillus foedidusin solid state cultures at different acidities. *Biotechnol. Lett.*, 18: 251-256.

Cherry, J. R., G. E. Nedwin and P. Falholt. 2001. Enzymes: Biotechnological Applications. In: Burlage, R. S. (Ed.). *The Encyclopedia of Environmental Microbiology*, John Wiley and Sons, Inc., New York. pp. 1176-1189.

Cochrane, V. W., 1958. *Physiology of the fungi.* John Wiley and Sons Inc., New York, New York.

Collins, P. J., Kotterman, M. J. J., Field, J. A., Dobson, A. D. W., 1996. Oxidation of anthracene and benzo[a]prene by laccase from Trametes versicolor. *Applied Environmental Microbiology* 62, 4563–4567.

Coughlan, M. P., Hazlewood, G. P. 1993. b-1, 4-D-Xylan-degrading enzyme systems: Biochemistry, molecular biology and applications. *Biotechnol. Appl. Biochem.*, 17:259-289.

Couto, S. R., Sanromán, M. Á., 2006. Effect of two wastes from groundnut processing on laccase production and dye decolourisation ability. *J. of Food Eng.* 73(4): 388-393.

Dashtban, M., Schraft, H. and Qin, W. (2009). Fungal Bioconversion of Lignocellulosic Residues; Opportunities and Perspectives. *Internetional Journal of Biological Sciences*, Vol. 5, No. 6, (September 2009), pp. 578-595, ISSN 1449-2288.

Dawe, A. L., Willins, D. A., et al. Increased transformation efficiency of *Aspergillus nidulans* protoplasts in the presence of dithiothreitol. *Anal. Biochem.* 2000; 283(1):111-2.

Dawson, L. and Boopathy, R. (2007). Use of Post-Harvest Sugarcane Residue for Ethanol Production. *Bioresource Technology*, Vol. 98, No. 9, (August 2006), pp. 1695-1699, ISSN: 0960-8524.

Dhillon, S. S., Gill, R. K., Gill, S. S., Singh, M., 2004. Studies on the utilization of citrus peel for pectinase production using fungus *Aspergillus niger*, *Intern. J. Environ. Studies*, Vol. 61(2), pp. 199-210.

Dhillon, G. S., Brar, S. K., Verma, D., and Tyagi, R. D. 2011. Utilization of different agro-industrial wastes for sustainable bioproduction of citric acid by *Aspergillus niger*. *Biochemical Engineering Journal* 54: 83-92.

Duarte, J. C., Pereira, J., Paixão, S. M., Baeta-Hall, L., Ribeiro, B., Sààgua, M. C., 2010. Ethanol production from enzymatically pretreated wheat straw. In: *Proceedings of 2nd International Conference of IAMAW (IAMAW'2010)*, Izmir, Turkey, June 17-19, 2010, (CD-ROM). F. V. Sukan, M. Raviv and N. Azbar (Eds). International association of Mediterranean Agro-industrial wastes (IAMAW). p. 89-94.

Duenas, R., R. P. Tengerdy and M. Giutierrez-Corria. 1995. Cellulase production by mixed fungi in solid substrate fermentation bagasse. *Biotechnol. Lett.* 8(5): 206-210.

Durrani, Y., S. Wahab and M. A. C. Javid-Ullah. 2002. Feutrell's early waste utilization in by-product development and its utilization in bakery products. *Sarhad J. Agric.* (Pakistan). 18(4): 463-466.

Dwivedi, P., V. Vivekanand, R. Ganguly and R. P. Singh, 2009. *Parthenium* sp. as a plant biomass for the production of alkalitolerant xylanase from mutant *Penicillium oxalicum* SAU_E-3.510 in submerged fermentation. *Biomass Bioenerg.*, 33: 581-588.

Ellaiah, P., Adinarayana, K., Bhavani, Y., Padmaja, P., and Srinivasula, B., 2002. Optimization of process parameters for glucoamylase production under solid state fermentation by a newly isolated *Aspergillus* species. *Process Biochemistry*, 38: 615-620.

Enguita, F. J., Martins, L. O., Henriques, A. O., and Carrondo, M. A. (2003) Crystal structure of a bacterial endospore coat component. A laccase with enhanced thermostability properties. *J. Biol. Chem.*, 278, 19416-19425.

Fadel, M., 2000. Production of thermostable amylolytic enzyme by Aspergillus niger F-909 under solid state fermentation. *Egypt. J. Microbiol.*, 35: 487-505.

Falcón, M. A., Rodríguez, A., Carnicero, A., et al. (1995). Isolation of microorganisms with lignin transformation potential from soil of Tenerife Island. *Soil Biol. Biochem.* 27(2): 121-126.

Fang, J. M., Fowler, P., Tomkinson, J., Hill, C. A. S. 2002. Preparation and characterization of methylated hemicelluloses from wheat straw. *Carbohyd. Polym.* 47: 285- 293.

Galante, Y., De Conti, A. and Monteverdi, R. 1998. Application of *Trichoderma* enzymes in the food and feed industries. 1998. In: Harman, G. and Kubicek, C. (eds.) *Trichoderma* and *Gliocladium, Enzymes, biological control and commercial applications*, Vol. 2. pp. 327-342.

Galembeck, F., Augusto Sales Barbosa, C. and Arromba De Sousa, R., 2009. Aproveitamento sustentável de biomassae de recursos naturais na inovação química. *Quim. Nova*, Vol. 32, No. 3, 571-581.

Gámez, S., González-Cabriales, J., Ramírez, J., Garrote, G., Vázquez, M., 2006. Study of the hydrolysis of sugarcane bagasse using phosphoric acid. *Journal of Food Engineering*, 74: 78-88.

Ghosh, B. and Ray, R. R. 2010. Saccharification of raw native starches by extracellular isoamylase of Rhizopus oryzae. *Biotechnologie* 9: 224-228.

Goes, A. P. and J. D. Sheppard. 1999. Effect of surfactants on alpha-amylase production in a solid substrate fermentation process. *Journal of Chemical Technology and Biotechnology*. 74(7): 709-712.

Grewal, H. S., Kalra, K. L., 1995. Fungal production of citric acid. *Biotechnol. Adv.* 13: 209-234.

Grohmann, K. and E. A. Baldwin. 1992. Hydrolysis of orange peel with pectinase and cellulase enzymes. *Biotechnol. Lett.* 14:1169-1174.

Gupta, R., Gigras, P., Mohapatra, H., Goswami, V. K., Chauhan, B., 2003. Microbial amylases: a biotechnological perspective, *Process. Biochem.*, 38:1599-1616.

Gupta, R., Beeg, Q. K., Loranz, P., 2002. Bacterial alkaline proteases: molecular approaches and industrial applications. *Appl. Microbiol. Biotechnol.* 59(1): 15-32.

Gutierrez-Correa, M., Portal, L., Moreno, P., Tengerdy, R. P., 1999. Mixed culture solid substrate fermentation of *Trichoderma reesei* with *Aspergillus niger* on sugarcane bagasse. *Biores. Technol.* 68:173-178.

Hagihara, H., K. Igarashi, Y. Hayashi, K. Endo and K. Ikawa-Kitayama, 2001. Novel α-amylase that is highly resistant to chelating reagents and chemical oxidants from the alkaliphilic Bacillus isolate KSM-K38. *Applied Environ. Microbiol.*, 67: 1744-1750.

Hara, T. and Ueda, S. 1982. Regulation of polyglutamate in *Bacillus subtilis* (natto): transformation of high PGA productivity. *Agric. Biol. Chem.* 46: 2275–2281.

Hashim, S. O., Delgado, O. D., Martı´nez, M. A., Kaul, R. H., Mulaa, F. J., Mattiasson, B. (2005). Alkaline active maltohexaose-forming aamylase from *Bacillus halodurans* LBK 34. *Enzyme Microb. Technol.* 36, 139-146.

Hublik, G., Schinner, F., 2000. Characterization and immobilization of the laccase from Pleurotus ostreatus and its use for continuous elimination of phenolic pollutants. *Enzyme and Microbial Technology* 27, 330-336.

Hynes, M. J., 1996. Genetic transformation of filamentous fungi. *J. Genet.*; 75(3): 297-311.

Imeri, A. G. and Knorr, D. 1998. Effects of chitosan on yield and compositional data of carrot and apple juice. *Journal of Food Science*, 53, 1707–1709.

Immanuel, G., C. M. A. Bhagavath, P. L. Raj, P. Esakkiraj and A. Palavesam. 2007. Production and partial purification of cellulase by *A. fumigates* fermented in coir waste and sawdust. *Internet J. Microbiol.* 3(1).

Javier, P. F. I., Ôscar, G., Sanz-Aparicio, J., Diaz, P. 2007. Xylanases: Molecular properties and applications. In: Polaina, J., MacCabe, A. P., eds. *Industrial Enzymes: Structure, Function and Applications*. Dordrecht, the Netherlands: Springer. 65-82.

Jiang, Z., Q. Cong, Q. Yan, N. Kumar and X. Du, 2010. Characterization of a thermostable xylanase from *Chaetomium* sp. and its application in Chinese steamed bread. *Food Chem.*, 120: 457-462.

Johannes, C., Majeherczyk, A., Hutterman, A., 1996. Degradation of anthracene by laccase of Trametes versicolor in the presence of different mediator compounds. *Applied Microbiology and Biotechnology* 46, 313–317.

Kadam, K. L., 1996. Cellulase production. In: Wyman, C. E., editors. *Handbook on bioethanol: production and utilization.* Washington, DC: Taylor and Francis. P. 213-252.

Kawai, S., Asukai, M., Ohya, N., Okita, K., Ito, T., Ohashi, H., 1999. Degradation of a non-phenolic b-O-4 substructure of Coriolus versicolor in the presence of 1-hydroxybenzotriazole. *FEMS Microbiology Letters* 170, 51–57.

Kennedy, M., List, D., Lu, Y., Foo, L. Y., Newman, R. H., Sims, I. M., Bain, P. J. S., Hamilton, B., Fenton, G., 1999. Apple pomace and products derived from apple pomace: uses, composition and analysis. *Modern Methods of Plant Analysis*, v. 20, p. 75-119.

Kim, S., Dale, B. E. (2004). Global Potential Bioethanol Production from Wasted Crops and Crop Residues. *Biomass and Bioenergy*, Vol. 26, pp. 361-375, ISSN 0961-9534.

Kinghorn, J. R., Hawkins, A. R., 1982. Cloning and expression in *Escherichia coli*-K-12 of the biosynthetic dehydroquinase function of the arom cluster gene from the eucaryote, *Aspergillus nidulans*. *Mol. Gen. Genet.*; 186(1):145-52.

Koutinas, A. A., Malbranque, F., Wang, R. H., Campbell, G. M., and Webb, C., 2007. Development of an oat-based biorefinery for the production of lactic acid by *Rhizopus oryzae* and various value-added co-products. *J. Agric. Food. Chem.* 55: 1755-1761.

Kumar, D., Jain, V. K., Shanker, G., Srivastava, A. (2003). Citric acid production by solid state fermentation using sugarcane bagasse. *Proc. Biochem.*, 38:1731-1738.

Lal, D. N., 1980. The influence of some ammonium compounds on the production of citric acid by *Aspergillus niger* AL. 29. *Ind. J. Agri. Chem.* 13: 153-157.

Latifian, M., Hamidi-Esfahani, Z. and Barzegar, M., 2007. Evaluation of culture conditions for cellulose production by two Trichoderma reesei mutants under solid-state fermentation conditions. *Bioresource Technology*, 98, 3634-3637.

Leangon, S., Maddox, I. S., Brooks, J. D., 2000. A proposed biochemical mechanism for citric acid accumulation by *Aspergillus niger*. Yang N 2 growing in solid state fermentation. *W. J. Microbiol. Biotechnol.* 16:271-275.

Lee, S.-K., George, S. D., Antholine, W. E., Hedman, B., Hodgson, K. O., and Solomon, E. I. (2002). Nature of the intermediate formed in the reduction of O_2 to H_20 at the trinuclear copper cluster active site in native laccase. *J. Am. Chem. Soc.*, 124, 6180-6193.

Leite, R. S. R., Alves-Prado, H. F., Cabral, H., Pagnocca, F. C., Gomes, E., Da-Silva, R. (2008). Production and Characteristics Comparison of Crude β-Glucosidases Produced by Microorganisms *Thermoascus aurantiacusand Aureobasidium pullulansin* Agricultural Wastes. *Enzyme and Microbial Technology*, Vol. 43, pp. 391–395, ISSN 0141-0229.

Lequart, C., Nuzillard, J. M., Kurek, B., Debeire, P., 1999. Hydrolysis of wheat bran and straw by an endoxylanase: production and structural characterization of cinnamoyl-oligosaccharides. *Carbohyd. Res.*; 319:102-111.

Ma, E., Cervera, Q., Mejía Sánchez, G. M., 1993. Integrated Utilization of Orange Peel, *Bioresource Technology* 44: 61-63.

Machius, M., Wiegand, G., Huber, R. 1995. Crystal structure of calcium-depleted Bacillus licheniformis α-amylase at 2. 2 Å. Resolutions, *J. Mol. Biol.*, 246, pp. 545–559.

Maciel, G. M., L. P. S. Vandenberghe, C. Windson, I. Haminiuk, R. C. Fendrich, B. E. D. Bianca, T. Q. D. Brandalize, A. Pandey and C. R. Soccol. 2008. Xylanase production by *Aspergillus niger* LPB 326 in solid-state fermentation using statistical experimental designs. *Food Technol. Biotechnol.* 46(2): 183-189.

Mamma, D., Kourtoglou, E., Christakopoulos, P., (2008). Fungal multienzyme production on industrial by-products of citrus-processing industry, *Bioresource Technology*, 99: 2373-2383.

Martins, E. S., Silvam, R. and Gomes, E. M 2000. Solid state production of thermostable pectinases from thermophilic *Thermoascus aurantiacus. Process Biochem.* 37: 949-954.

Mayer, J., Wiley, B., Henderson, K., and Kaplan, D. 1989. Physical properties of films produced from the biopolymers pollulan and chitosan produced by *Aureobasidicum pullulans* and *Mucor rouxii* Abstracts, *annual meeting of the American society for microbiology.* Washington, DC: ASM.

Miettinen-Oinonen, A., 2004. *Trichoderma reesei* strains for production of cellulases for the textile industry. Otamedia Oy, Espoo 2004. VTT Publications 550. 96 p. + app. 53 p. http://ethesis.helsinki.fi/julkaisut/bio/bioja/vk/miettinen-oinonen/trichode.pdf.

Mirzaei-Aghsaghali, A. and N. Maheri-Sis, 2008. Nutritive value of some agro-industrial by-product for ruminants-a review. *World J. Zool.*, 3: 40-46.

Mishra, A., Kumar, S., 2007. Cyanobacterial biomass as N-supplement to agro-waste for hyper-production of laccase from Pleurotus ostreatus in solid state fermentation. *Process Biochem.* 42(4): 681-685.

Muzzarelli, R. A. A., 1989. Amphoteric derivatives of chitosan and their biological significance. In: G. Skjak-Braek, T. Anthonsen and P. Sandford (Eds.), *Chitin and chitosan* (pp. 87–99). London: Elsevier.

Nandakumar, M. P., Thakur, M. S., Raghavarao, K. S. M. S., Ghildyal, N. P. (1994). Mechanism of solid particle degradation by Aspergillus nigerin solid state fermentation. *Proc. Biochem.*, 29:545-551.

Napolitano, A., Lanzuise, S., Ruocco, M., Arlotti, G., Ranieri, R., Knutsen, S. H., Lorito, M., Fogliano, V. 2006. Treatment of cereal products with a tailored preparation of Trichoderma enzyme increases the amount of soluble dietary fiber. *Journal of Agriculture and Food Chemistry*, 54:7863–7869.

Nierstrasz, V. and Warmoeskerken, M. 2003. Process engineering and industrial enzyme applications. In: Cavaco-Paulo, A. and G¸bitz, G. (eds.) *Textile processing with enzymes*, Woodhead Publishing Ltd., England. pp. 120-157.

Nigam, P. and Singh, D., 1995. Enzymes and microbial enzymes involved in starch processing enzymes, *Microb. Technol.* 17: 770-778.

Onishi, H., Nagai, T., Machida, Y. (1997). Application of chitin, chitosan,and their derivatives to drug carriers for microparticulated or conjugated drug delivery systems. In: M. F. A. Goosen (Ed.), *Applications of chitin and chitosan* pp. 205–231. Lancaster, PA: Technomic Publishing Company.

Pandey, A., 2003. Solid state fermentation. *Biochem. Eng. J.;* 13:81–4.

Pandey, A., Soccol, C. R., Nigam, P., Soccol, V. T., Vandenberghe, L. P. S., and Mohan, R. (2000). Biotechnological potential of agro-industrial residues. II: cassava bagasse. *Bioresource Technology*, 74, 81-87.

Parameswaran, B. (2009). Sugarcane Bagasse. In: Nigam, P., Pandey, A. (editors) *Biotechnology for Agro-Industrial Residues Utilisation*. Springer, Netherlands.

Pelaez, F., Martínez, M. J., Martínez, A. T. (1995) Screening of 68 species basidiomycetes for enzymes involved in lignin degradation. *Mycol. Res.* 99: 37-42.

Penttilä, M., Limon, C. and Nevalainen, H. 2004. Molecular biology of *Trichoderma* and biotechnological applications In: Arora, D. (ed.) *Handbook of fungal biotechnology*. Marcel Dekker, Inc. Pp. 413-427.

Philippidis, G. P., 1994. Cellulose production technology-evaluation of current status. In: Himmel, M. E., Baker, J. O., Overend, R. P., editors. *Enzymatic conversion of biomass for fuels production*. Washington, DC: ACS.

Rein, P., 2007. *Cane Sugar Engineering*. Berlin: Verlag Dr. Albert Bartens K. G. Berlin, 768 p.

Reinhammar, B., 1984. Laccase. In: *Copper proteins and copper enzymes*, Vol. 3. (Lontei, R., Ed.) CRC Press, Boca Raton. p. 1-35.

Robinson, T., Singh, D., Nigam, P. 2001. Solid-state fermentation: a promising microbial technology for secondary metabolites production. *App. Microbiol. Biotechnol.* 55:284-289.

Rodríguez-Couto, S., 2008. Exploitation of biological wastes for the production of value-added products under solid-state fermentation conditions. *Biotechnol. J.* 3, 859–870.

Romanowska, I., J. Polak and S. Bielecki, 2006. Isolation and properties of *Aspergillus niger* IBT-90 xylanase for bakery. *Applied Microbiol. Biotechnol.*, 69: 665-671.

Sajitha, N., Vasanthabharathi, V., Lakshminarayanan, R., and Jayalakshmi, S. (2011), "Amylase from an Estuarine Bacillus megaterium", *Current Research Journal of Biological Sciences,* 3 (2), 110-115.

Sanchez, O. and Aguirre, J., 1996. Efficient transformation of Aspergillus nidulans by electroporation of germinated conidia, *Fungal Genetics Newsletter*, 43 (1996), pp. 48–51.

Sapag, A., Wouters, J., Lambert, C., Ioannes, P., Eyzaguirre, J., and Depiereux, E., The endoxylanases from family 11: computer analysis of protein sequences reveals important structural and phylogenetic relationships, *J. Biotechnol.* (In Press), 2002.

Sasaki, M., Adschiri, T., Kunio, A. (2003), Fractionation of sugarcane bagasse by hydrothermal treatment. *Bioresource Technology*, 86: 301-304.

Sato Mariana Fátima; Renato Giovanetti Vieira; Danianni Marinho Zardo; Leila Denise Falcão; Alessandro Nogueira; Gilvan Wosiacki. 2010. Apple pomace from eleven cultivars: an approach to identify sources of bioactive compounds. *Acta Sci., Agron.* Vol. 32 no.1 Maringá.

Schallmey, M., Singh, A., Ward, O. P. 2004. Development in the use of *Bacillus* species for industrial production. *Can. J. Microbiol.*, 50, 1-17.

Schieber, A., Hilta, P., Strekera, P., Ulrich, H., Rentschlerb, C., and Carle, R. 2003. A new process for the combined recovery of pectin and phenolic compounds from apple pomace. *Innovative Food Science and Emerging Technologies*, 4(1), 99-107.

Shah, A. R. and D. Madamwar, 2005. Xylanase production by a newly isolated *Aspergillus foetidus* strain and its characterization. *Process Biochem.,* 40: 1763-1771.

Shalani, R. and Gupta, D. K. 2010. Utilization of pomace from apple processing industries: a review. *Journal of Food Science and Technology*, 47(4), 365-371.

Silva, D., E. S. Martins, R. Silva and E. Gomes, 2002. Pectinase production from *Penicillium viridicatum* RFC3 by solid state fermentation using agricultural residues and agro-industrial by-product. *Braz. J. Microbiol.*, 33: 318-324.

Smock, R. M. Neubert, A. M., 1950. *Apples and Apple Products*. Interscience Publishers: New York, 1950. p. 486.

Solange I. Mussatto, Lina F. Ballesteros, Silvia Martins and José A. Teixeira, 2012. *Use of Agro-Industrial Wastes in Solid-State Fermentation Processes, Industrial Waste*, Prof. Kuan-Yeow Show (Ed.), ISBN: 978-953-51-0253-3, InTech, Available from: http://www.intechopen.com/books/industrial waste use of agro industrialwastes in solid state fermentation processes.

Srebotnik, E., Hammel, K. E., 2000. Degradation of nonphenolic lignin by the laccase/1-hydroxybenzotriazole system. *Journal of Biotechnology* 81, 179-188.

Subramaniyan, S., Prema, P., 2002. Biotechnology of Microbial Xylanases: Enzymology, Molecular Biology and Application. *Critical Reviews in Biotechnology*. Vol. 22 (1), pp 33-46.

Synowiecki, J. and Al-Khateeb, N. A. 1997. Mycelia of *Mucor rouxii* as a source of chitin and chitosan. *Food Chemistry*, 60, 605- 610.

Tengerdy, R. P. and Szakacs, G., 2003. Bioconversion of Lignocellulose in Solid Substrate Fermentation. *Biochemistry and Engineering Journal*, Vol. 13, pp. 169-179, ISSN 1369-703X.

Torres, E., Bustos-Jaimes, I., Le Borgne, S., 2003. Potencial Use of Oxidative Enzymes for the Detoxification of Organic Pollutants. *Applied Catalysis Environmental*, Vol. 46, pp. 1-15, ISSN 0926-3373.

Ueda, S. 1989. Industrial applications of *B. subtilis*: utilization of soybean as natto, a traditional Japanese food. *In: Bacillus subtilis: molecular biology and industrial applications*. Edited by B. Maruo and H. Yoshikawa. Elsevier Science B.V., Amsterdam, Netherlands. pp. 143-161.

UNSD (United Nations Statistics Division) (2005). Energy Statistics Database. Retrieved October 2012 from http://www.nationmaster.com/graph.

Van der Maarel, M. J. E. C., B. van der Veen, J. C. M. Uitdehaag, H. Leemhuis and L. Dijkhuizen, 2002. Properties and applications of starch converting enzymes of the α-amylase family. *J. Biotechnol.*, 94: 137-155.

Venus, J. and Richter, K., 2006. Production of lactic acid from barley: Strain selection, phenotypic and medium optimization. *Eng. Life Sci.* 6(5): 492-500.

Vibha Bhardwaj, Neelam Garg, 2011. The utilization of citrus peel for pectinase production. *Journal of Sustainable Development and Environmental Protection* Vol. 1 No.3; 60-66.

Vicuña, R., 1988. Bacterial degradation of lignin. *Enzyme Microb. Technol.* 10:646-655.

Virk, B. S., Sogi, D. S., 2004. Extraction and characterization of pectin from apple (Malus pumila) peel waste. *International Journal of Food Properties*, Vol. 7, N. 3, p. 693-703.

Wan, M. W., Petrisor, I. G., Lai, H.-T., Kim, D., and Yen, T. F., 2004. Copper and sorption through chitosan immobilized on sand to demonstrate the feasibility for in situ soil decontamination. *Carbohydrate Polymers*, 55, 249-254.

Wang, B. D., D. C. Chen and T. T. Kuo, 2001. Characterization of a *Saccharomyces cerevisiae* mutant with over secretion phenotype. *Applied Microbiol. Biotechnol.*, 55: 712-720.

Wang, L.-F. and Doi, R. H. 1989. Heterologous gene expression in Bacillus subtilis. In: *Biology of bacilli: applications to industry.* Edited by R. H. Doi and M. McLaughlin. Butterworth-Heinemann, Boston, Mass. pp. 63–104.

Wong, K. K. Y., Tan, L. U. L. and Saddler, J. N., 1988. Multiplicity of β-1, 4-xylanase in microorganisms: Functions and applications, *Microbiol. Rev.* 52, 305.

Wongwicharn, A., Harvey, L. M., McNeil, B., 1999. Secretion of heterologous and native proteins, growth and morphology in batch cultures of *Aspergillus niger* B1-D at varying agitation rates. *J. Chem. Technol. Biotechnol.* 74: 821-828.

Wosiacki, G., Sataque, E. Y., 1987. Caracterização da polifenoloxidase da maçã (Malus domestica, variedade Gala). *Arquivos de Biologia e Tecnologia*, v. 30, n. 2, p. 287-299.

Xu, J. F., Wang, L. P., Ridgway, D., Gu, T. Y., Moo-Young, M., 2000. Increased heterologous protein production in *Aspergillus niger* fermentation through extracellular proteases inhibition by pellet growth. *Biotechnol. Prog.* 16: 222-227.

Yu, J., Stahl, H., 2008. Microbial utilization and biopolyester synthesis of bagasse hydrolysates. *Bioresource Technology*, 99: 8042-8048.

Zimmermann, W., 1990. Degradation of lignin by bacteria. *J. Biotechnol.* 13(2-3):119-130.

In: Enzymes in Value-Addition of Wastes
Editors: Satinder Kaur Brar and Mausam Verma
ISBN: 978-1-62948-968-1

Chapter 7

HYDROLYSIS OF CASSAVA BAGASSE STARCH USING AMYLASES FOR VALUE-ADDITION

***Rojan P. John*[1], *G. S. Anisha* [2] *and K. Madhavan Nampoothiri*[3]**
[1]GHSS Pookkottumpadam, Malappuram, Kerala, India
[2]Department of Zoology, Govt. College for Women, Thiruvananthapuram, Kerala, India
[3]Biotechnology Division, National Institute for Interdisciplinary Science & Technology, Thiruvananthapuram, Kerala, India

ABSTRACT

Cassava (*Manihot esculenta* Cranz), root crop, is the third most important source of calories in the tropics, after rice and corn. Industrial processing of cassava tubers to isolate starch generates more liquid and solid residues. Solid residues include brown peel, inner peel, unusable roots, bagasse and flour refuse, among which bagasse is the main residue. Cassava bagasse is made up of fibrous root material and contains starch that could not be extracted by physical process. Poor processing conditions may result in even higher concentrations of starch in cassava bagasse. Cassava bagasse can be used for the production of value added products. Hydrolysis of cassava bagasse gives fermentable sugars and can be used as a liquid medium for microbial growth. Enzymatic hydrolysate of cassava bagasse is best medium for microbial fermentation compared to hydrolysate by acid hydrolysis. In the present study, optimization of liquefaction and saccharification of gelatinized cassava bagasse were carried out using response surface methodology. Cassava bagasse (75 g/L) liquefied with 2000 U glucoamylase for 60-70 min gave more than 40 g/L reducing sugar and provided the maximum conversion than other concentrations of cassava bagasse.

INTRODUCTION

Cassava (*Manihot esculenta* Crantz) is a short-lived perennial belonging to Euphorbiaceae. The origin of cassava is in eastern Brazil of South America. Cultivation is usually by stem cuttings, and each plant produces very fleshy adventitious roots up to 15 centimeters in diameter. Cassava is the third most important source of calories in the tropics, after rice and corn [1]. More than 600 million people in Africa, Asia and Latin America depend on the cassava [2]. Total cassava production is estimated as approximately 166 million tons and is projected to reach approximately 266 million tons by 2020. Most of the contribution in the production is from the developing countries and from the sub-Saharan Africa [3]. Industrial processing of cassava tubers is mainly done to make flour and isolate starch, which generates more liquid and solid residues (processing for flour generates solid residues while for starch generates more liquid residues besides solid bagasse and other wastes). Solid residues generated from cassava tuber processing consist of outer thin brown peel, inner fleshy peel, unusable roots and bagasse and bagasse is the main residue among all others. Surprisingly, during processing about 250–300 t of fresh tubers results about 280 t of wet cassava bagasse.

Cassava bagasse contains starch entrapped in fibrous matter and that could not be extracted by the physical separation. Concentration of starch may vary depending upon the processing conditions. Concentrations of starch in cassava bagasse may be increased due to poor processing conditions.

The starch has to be recovered in the form of polymeric or monomeric forms to reduce the environmental problems. Biotechnological or physico-chemical methods can be applied to recover these sugars. Improved starch recovery from cassava bagasse can be achieved by using enzyme preparations. In current study, amylase enzymes were utilized to extract sugars from cassava bagasse and the process was optimized using response surface method.

MATERIALS AND METHODS

Enzymes

For the enzymatic hydrolysis of cassava bagasse, commercially available (Rashesh and Co., Mumbai, India) α-amylase (Termamyl) and glucoamylase (AMG) were used.

Cassava Bagasse and Hydrolysis

Cassava bagasse was collected from Varalakshmi starch industries, Selam, Tamil Nadu, India. Unless indicated, gelatinized cassava bagasse (7.5 %, w/v) was for hydrolysis.

Different concentrations of cassava bagasse and enzymes for liquefaction (Table 1) and saccharification (Table 3) were statistically optimized using Box-Behnken design. Starch concentration was estimated by determining the liberated glucose by acid hydrolysis and reducing sugar was determined by DNS method [4] as described previously.

RESULTS AND DISCUSSION

Cassava bagasse contains a large amount of starch trapped in the fibrous matrix and it is used for the production of many value added products. In the present study it contains ~52% starch. In order to optimize the process,the optimization of the conditions required for making bagasse hydrolysate, which contains reducing sugar as the sole carbon source, required for production of other fermentative products. Amount of cassava bagasse required for the proper gelatinisation, liquefaction and saccharification were tested experimentally.

Cassava Bagasse Hydrolysis

Liquefaction

Response surface optimization is more advantageous than the traditional single parameter optimization in that it saves time, space and raw material. The current design was applied to the hydrolysis of cassava bagasse starch by liquefaction. The design used and results of liquefaction optimization are shown in Table 1.

Table 1. Box-Behnken design matrix for optimization of process variables for cassava bagasse liquefaction and the resultant reducing sugar concentration

Run	Variables			Reducing sugar (g/L)[a]
	Cassava bagasse (g/L)	Liquefaction duration (min)	Enzyme concentration (U/L)	
1	50	20	571	10.8
2	75	20	1000	17.6
3	75	10	1429	15.9
4	50	30	1000	11.2
5	100	20	571	20.7
6	100	10	1000	19.87
7	50	20	1429	10.6
8	75	20	1000	18.9
9	50	10	1000	11.17
10	75	30	1429	19.95
11	75	10	571	13.81
12	75	20	1000	19.78
13	75	20	1000	19.96
14	100	30	1000	21.28
15	100	20	1428	21.35
16	75	20	1000	19.32
17	75	30	571	19.79

[a] Values are mean of three replicate experiments.

The data was analyzed by multiple regression analysis using the Design Expert software and the following polynomial equation was derived to represent lactic acid yield as a function of the independent variables tested for liquefaction. The results of the study showed that the cassava bagasse concentration in the medium was the most significant single parameter which influenced liquefaction followed by duration and enzyme concentration (Table 4 & Equation 1).

$$Y = 19.1 + 5 X_1 + 1.42 X_2 + 0.4 X_3 - 2.36 X_1^2 - 0.88 X_2^2 - 0.89 X_3^2 + 0.35 X_1 X_2 + 0.21 X_1 X_3 - 0.5 X_2 X_3 \quad \text{Eq. (1)}$$

Where Y is the predicted reducing sugar yield from liquefaction and x_1, x_2 and x_3 are the coded values for cassava bagasse, liquefaction duration and enzyme concentration respectively.

The experimental data was statistically analyzed using the Fischer's statistical test for analysis of variance (ANOVA) and the results are shown in Table 2.

Table 2. Analysis of variance (ANOVA) for the response surface quadratic model for cassava bagasse liquefaction

Source	Prob > F	Significance
Model	0.0011	☑
X_1	< 0.0001	☑
X_2	0.0236	☑
X_3	0.5193	☒
X_1^2	0.0100	☑
X_2^2	0.2298	☒
X_3^2	0.2218	☒
$X_1 X_2$	0.6389	☒
$X_1 X_3$	0.7714	☒
$X_2 X_3$	0.5149	☒
Lack of Fit	0.1269	☒

☑ Significant variable, ☒ Insignificant variable.

The ANOVA of the quadratic regression model indicated that the model was highly significant. The *Prob*>F value of the model was 0.0011, which also confirmed that the model was highly significant. The coefficient estimate and the corresponding *Prob*>F values suggested that all the independent variables studied had a significant effect on liquefaction. The interactions of selected parameters are shown in contour graph (Figure 1).

Saccharification

From the liquefaction results, a range of cassava bagasse concentration (75-105 g, liquefied with 133 U glucoamylase/g cassava bagasse for 25 min) has taken for the saccharification optimization. The design for saccharification optimization and the yield obtained is shown in Table 3. The results of the study showed that the duration was the most significant single parameter which influenced saccharification (Table 4 & Equation 2).

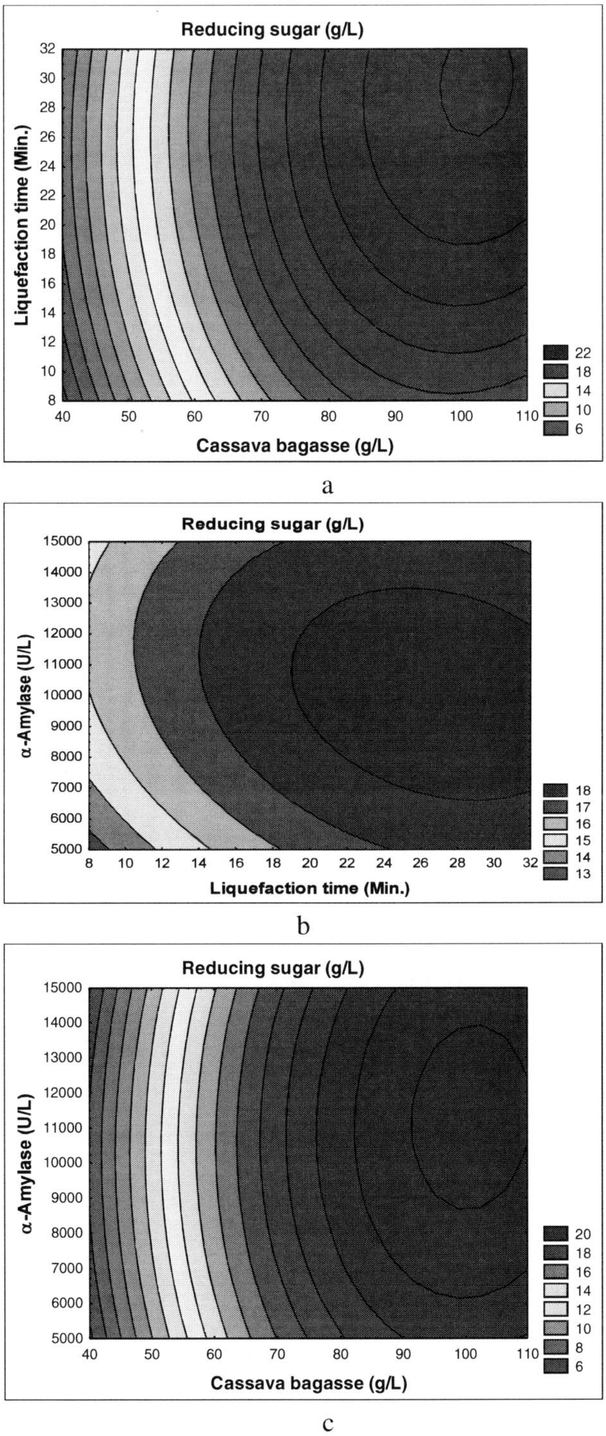

Figure 1. Interactions of (a) duration and cassava bagasse concentration, (b) duration and amylase concentration and (c) cassava bagasse and amylase concentration on liquefaction of cassava bagasse.

$$Y = 38.22 + 9.86\,X_1 + 2.85\,X_2 + 0.695\,X_3 - 4.71\,X_1^2 - 1.75\,X_2^2 - 1.79\,X_3^2 + 0.69\,X_1X_2 + 0.43\,X_1X_3 - 0.93\,X_2X_3 \qquad \text{Eq. (2)}$$

Where Y is the predicted reducing sugar yield from liquefaction and x_1, x_2 and x_3 are the coded values for saccharifying duration, enzyme concentration and cassava bagasse respectively.

Table 3. Box-Behnken design matrix for optimization of process variables for cassava bagasse saccharification and the resultant reducing sugar concentration

Run	Variables			Reducing sugar
	Saccharifying duration	Enzyme	Cassava bagasse concentration	
	(min)	(U/L)	(g/L)	(g/L)[a]
1	30	2000	75	21.6
2	50	2000	90	35.2
3	50	1140	105	31.8
4	30	2860	90	22.4
5	70	2000	75	41.4
6	70	1140	90	39.74
7	30	2000	105	21.2
8	50	2000	90	37.8
9	30	1140	90	22.34
10	50	2860	105	39.9
11	50	1140	75	27.62
12	50	2000	90	39.56
13	50	2000	90	39.92
14	70	2860	90	42.56
15	70	2000	105	42.7
16	50	2000	90	38.64
17	50	2860	75	39.42

[a] Values are mean of three replicate experiments.

Table 4. Analysis of variance (ANOVA) for the response surface quadratic model for cassava bagasse saccharification

Source	Prob > F	Significance
Model	0.001	☑
X_1	< 0.0001	☑
X_2	0.0229	☑
X_3	0.5015	☒
X_1^2	0.0102	☑
X_2^2	0.2362	☒
X_3^2	0.2278	☒
AB	0.6342	☒
AC	0.7682	☒
BC	0.5263	☒
Lack of Fit	0.1175	☒

☑ Significant variable, ☒ Insignificant variable.

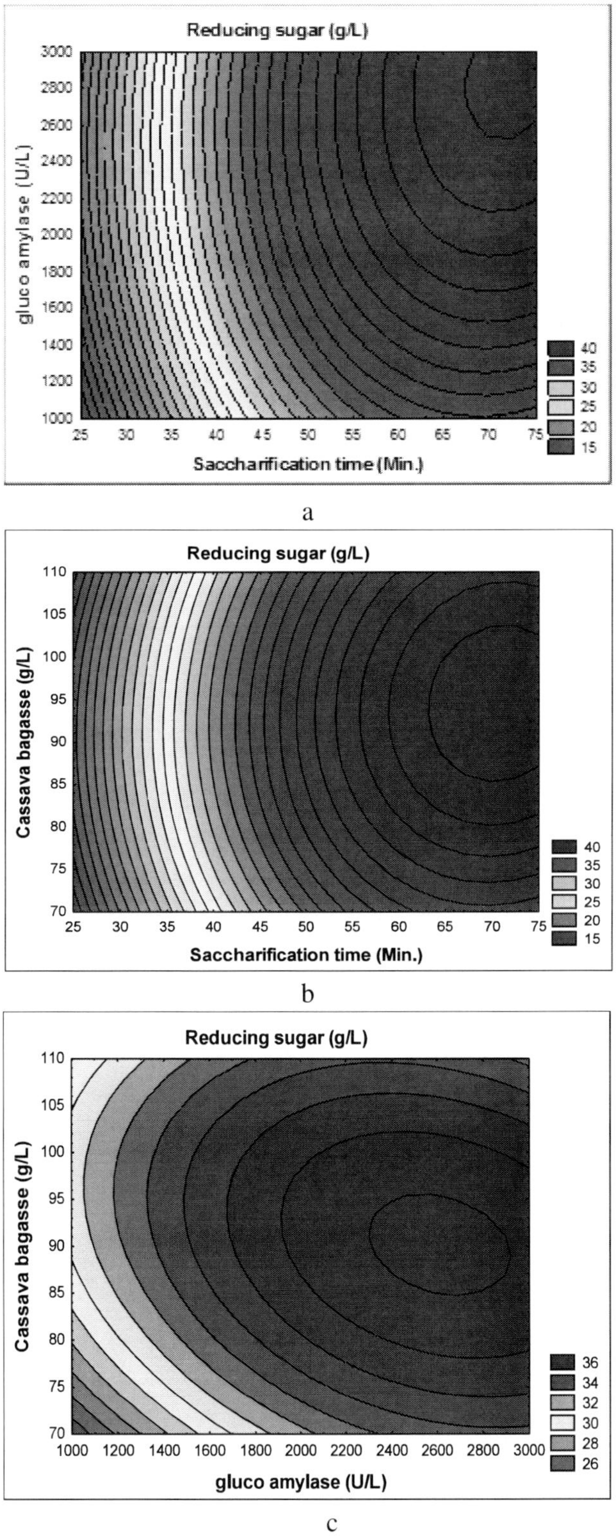

Figure 2. Interactions of (a) duration and glucoamylase concentration, (b) duration and cassava bagasse concentration and (c) cassava bagasse and glucoamylase concentrations on saccharification.

The interactions of selected parameters were shown in contour graph (Figure 2). Cassava bagasse (75 g/L) liquefied with 2000 U glucamylase for 60-70 min gave more than 40 g/L reducing sugar and has the maximum conversion than other concentrations of cassava bagasse. For the further studies 7.5% cassava bagasse treated under optimized conditions to yield maximum hydrolysis.

CONCLUSION

The shake flask hydrolysis showed that 7.5% (w/v) cassava bagasse gave the best conversion and a hydrolysate containing more than 40 mg/mL of reducing sugar was obtained after the liquefaction and saccharification process. The difficulty of solidification during gelatinization was noticed during hydrolysis if amount of cassava bagasse was higher than 7.5% (w/v). For further experiments, 7.5% (w/v) cassava bagasse was selected. The amount of α-amylase and glucoamylase and the time required for the liquefaction and saccharification were also optimized. About 1000 IU of α-amylase at 90 °C for 30 min was the optimum for liquefaction and 200 IU of glucoamylase at 60 °C for 70 min for saccharification. More than 95% of starch was converted to the reducing sugar in the process conditions. The hydrolysate obtained from enzymatic hydrolysis of cassava bagasse was effectively utilized for the production of lactic acid by submerged and solid state fermentation [5-8].

REFERENCES

[1] John, R.P. Biotechnological Potentials of Cassava Bagasse. P. Singh nee' Nigam, A. Pandey (eds.), *Biotechnology for Agro-Industrial Residues Utilisation*, DOI 10.1007/978-1-4020-9942-7 11,Springer Science+Business Media B.V. 2009, pp. 225-237(2009).
[2] Elkholy, H. and Eltantawy, A. *J. Root Crops,* 26, 1 (2000).
[3] Tonukari, N.J. *Electronic J. Biotechnol.*, 7, 5 (2004).
[4] Miller, G.L. (1959) Use of dinitrosalicylic acid reagent for determination of reducing sugar. *Anal. Chem.,* 31, 426-429.
[5] Rojan, P.J. Nampoothiri, K.M. Nair, A.S and Pandey, A. *Biotechnol. Lett.*, 27, 1685 (2005).
[6] John, R.P. Nampoothiri, K.M and Pandey, A.. *Proc. Biochem.*, 41, 759 (2006).
[7] John, R.P. Nampoothiri, K.M and Pandey, A. *J. Basic Microbiol.*, 47, 25 (2007a).
[8] John, R.P. Nampoothiri, K.M and Pandey, A. *Lett. Appl. Microbiol.*, 44, 582 (2007b).

In: Enzymes in Value-Addition of Wastes
Editors: Satinder Kaur Brar and Mausam Verma
ISBN: 978-1-62948-968-1

Chapter 8

XYLANASES: ENZYMES FOR PLANT BIOMASS TREATMENT

Rojan P. John[1] and G. S. Anisha[2]
[1]GHSS Pookkottumpadam, Malappuram, Kerala, India
[2]Department of Zoology, Govt. College for Women, Thiruvananthapuram, Kerala, India

ABSTRACT

Xylanases is a class of enzymes which degrade the xylan molecules. Xylan is the most abundant non-cellulosic natural polymer present in hemicellulose of plant cell wall. Due to the complexity of xylan molecules (heteropolymers), a cocktail of enzymes is needed for its complete hydrolysis. Thus, xylanases comprises endo-1,4-β-xylanases (hydrolyze xylan randomly), 1,4-β-D-xylosidases (hydrolyse oligosaccharides), and α-L-arabinofuranosidase, α-D-glucuronidase, galactosidase and acetyl xylan esterase (remove side groups present in xylan). Xylanases are a major group of commercial enzymes having major applications in pulp industry as they hydrolyse xylan and release lignin from paper pulp and which reduce usage of chlorine as the bleaching agent. The agro-industrial residues can be treated with these enzymes to generate fermentable sugars for value addition. Bacteria including actinomycetes and fungi are considered as the potent producers of xylanase at commercial level. Agricultural residues can be used as low-cost substrates for the production of industrial enzymes and is a significant way to reduce production cost. Different agroindustrial residues are used to produce xylanase for industrial applications. This chapter reviews microbial production and applications of xylanases.

INTRODUCTION

Xylan is considered as the most abundant natural polymer next to cellulose. Xylan is one of the major components in cell wall material (middle lamellae) and is a complex polysaccharide in which predominant is β-D-xylose units. Xylans are classified on the basis of linkages to xylose residues. The β-1,3-linked xylans (found only in marine algae), containing a mixture of β-1,3- and β-1,4-linkages and β-1,4-linked xylans occur in hardwoods, softwoods and grasses [1]. Xylan content is more in angiospermic hardwood plant (15-30%) and in annual plants (<30%), than in gymnospermic soft woods (7-10%) [2]. Due to the complex nature, xylan has less economic value and is an antinutritional factor of plant biomass. The plant biomass including agro-industrial residues can be hydrolyzed to get easily digestible/accessible nutrient rich substrate for value addition and feed for poultry and cattle. Due to the abundant nature of this polymer, it should be treated or value added to reduce the problem of disposal.

Even though it is economically feasible, acid hydrolysis of xylan is not acceptable due to the production of toxic components during hydrolysis. The biotechnological conversion of complex molecules to easily accessible molecules has gained worldwide acceptance due to its eco-friendly nature.

Xylanases are hydrolytic enzymes coming under the family of glycoside hydrolase. Even though most of these enzymes belong to families 10 and 11, there are reports on xylanases belong to families 5,7,8 and 43 [2]. There is a need of cocktail of xylanase enzyme for the complete hydrolysis of highly complex and heterogenic xylan. Xylans of different plant origin contain varied amounts of arabinose, 4-*O*-methylglucuronic acid and acetic acid groups attached to the xylose chain back bone [1, 3-4]. Endo- 1,4- β-D-xylanase can cleave the backbone of substrate randomly. The substrates of β-D-xylosidase include xylooligosaccharides and xylobiose and which cut the monomeric unit from the non-reducing end. The α-L-Arabinofuranoxylanosidase, α-D-glucuronidase, acetyl xylan esterases, ferulic acid esterases, *p*-coumaric acid esterases etc. remove side group of xylan complex. Microorganisms, such as heterogenic saprobes/parasites produce these enzymes for getting xylose as their carbon source for metabolic activity or to cleave/penetrate the host body for infection. For commercial treatment of agro-industrial wastes there is a need of synergism of enzymes. Side chains cleaving enzymes help to access backbone cleaving enzymes by hydrolyzing the side chains.

Microbial Sources of Xylanases

Last few decades witnessed high growth rate in production of commercial enzymes and many works were carried out for the isolation of xylanase producing microorganisms for industrial production and applications. Both bacterial and fungal strains were used to produce xylanases. The microorganisms used for the production of these enzymes were *Aspergillus niger*, *Fusarium* sp., *Trichoderma* sp., *Penicillium* sp., *Streptomyces* spp. and *Bacillus* spp. [5-9]. Important strains used for the production of xylanases are shown in Table. 1. Bacterial cultures are preferred to fungal cultures as it has xylanases of higher pH optima than fungal enzymes. Fungal enzymes contain large quantity cellulase and are not preferable for the application where no cellulose cleavage is desirable. During scale up process the fragile fungal mycelium is disrupted by agitation in bioreactor and hence influences the enzyme production [10]. But few exceptions, such as *Thermomyces lanuginosus* are able to produce cellulase-poor/free xylanase using lignocellulosic materials [11-12]. Khucharoenphaisan et al. [13] studied the characteristics of purified enzymes from strains of *T. lanuginosus* (THKU-9 and THKU-49) and half-lives at 70 °C in 50 mM phosphate buffer (pH 6.0) were 178 and 336 min, respectively. Fungal cultures are preferred for commercial production as certain bacterial strains need anaerobic conditions [14]. Alkali tolerant, cellulase free, thermostable enzymes are preferred for certain applications. Thermostable xylanases were produced from hyperthermophilic bacterium *Thermotoga maritima* MSB8 [15]. Sharma et al. [16] reported alkali-thermostable and cellulase free xylanase production by an extremothermophile *Geobacillus thermoleovorans*. Site directed mutagenesis was carried out for improvement of thermostability [17, 18]. Subramaniyan, [19] reported that *Bacillus* sp. SSP34 produced xylanase with negligible amount of cellulase and protease on optimized medium. Protease-free enzyme mixture is desired as it has more stability since there is no enzyme degradation by the proteolytic enzymes. Maintaining easily available nitrogen in the medium for the growth of microorganisms limits the secretion of protease. But this character varies from microorganisms to microorganisms. Smith & Wood, [20] developed a medium for xylanase production by *Aspergillus awamori* while maintaining low protesase concentration.

On the other hand the saccharification of agroindustrial residues and fruit juice processing using enzyme mixture containing xylanase along with cellulase have prime position as it improve the digestibility of biomass [21]. The enzyme production capacity of organisms varies from organisms to organisms and substrate to substrate. One of the *Bacillus* spp. produced 506 IU/ml xylanase [19] and *B. circulance* produced 400 IU/ml [22]. While *Phanerochaete chrysosporium* produced ~20 U/ml of xylanase and ~2.5 U/ml cellulases [23], *Schizophyllum commune* produced 1244 U/ml xylanase and obvious existence of cellulase [24].

Single microorganism able to produce more than one form of xylanase (multiple forms) such as *Aspergillus niger* (15 xylanases), *Trichoderma viride* (13 forms) and *Phanerochaete chrysosporium* (more than 30 forms) are also reported [10].

Xylanase Production Using Agroindustrial Residues

Pure substrates, such as xylan is expensive to be used for bulk industrial production of xylanases [28]. Different lignocellulosic agroindustrial residues are tested for the growth of xylanase producing organisms. Both bacteria and fungi were grown in liquid and solid medium to optimize the production parameters. Archana & Satyanarayana [35] reported xylanase production by thermophilic *Bacillus licheniformis* A99 in solid state fermentation. Submerged and solid-state fermentation using *Penicillium canescens* 10-10c were tested for the use of agroresidue such as soya meal and wheat straw [27, 36]. Priem et al. [37] reported production of β-1,4-xylanase in continuous culture by *Aureobasidium pullulans* CBS 58475. Assamoi et al. 2008 reported that the maximal enzymatic production using *Penicillium canescens* 10-10c in soy oil cake based medium attained after 7 days incubation was 14 485 $U.g^{-1}$ of substrate in Erlenmeyer. According to them not only the cultivation conditions but also composition of the medium, cultivation temperature, induction and repression mechanisms and nature of inoculum are major causes for maximum yield. Dai et al. [8] optimized the process parameters for the production of xylanase A using orange peel as a substrate for *A. niger* and found that concentrations of the added glycerin and ammonium sulfate, the moisture content, and fermentation time had major role on yield. Wheat bran has xylan as the major poly saccharide and which is cheap and has universal generation. Yang et al. [38] and Bajaj et al. [28] studied the use of this residue for the production of xylanase using bacterial culture and reported the production of xylanase active at pH 6-8 and temperature 40-50°C. According to Bajaj et al. [28] most simpler and complex nitrogen sources has lesser effect on xylanase production while using *B. pumilus* SS1 in wheat bran medium. But certain other organisms prefer complex nitrogen sources for maximum production of xylanase [39].

Table 1. Important microorganisms used for production of xylanase

Microorganism	Bacteria/Fungi	Substrate used	Reference
Bacillus spp.	Bacteria	Wheat straw, wheat bran and corncob	[6]
Aspergillus niger	Fungus	Wheat bran, rice bran, ragi bran, soya bran and saw dust	[7]
A. niger and *Fusarium oxysporum*	Fungi	Forest wastes (*Toona ciliata, Celtris australis, Cedrus deodara* and *Pinus roxburghii)*	[9]
A. niger JL-15	Fungus	Orange peel	[8]
A. Niger SS7	Fungus	Agricultural residues such as corn cob hull	[25]
Neocallimastix frontalis	Fungus	Barley straw	[26]
Thermomyces lanuginosus	Fungus	Synthetic medium	[13]
Penicillium canescens	Fungus	Soy oil cake	[27]
Bacillus pumilus SS1	Bacterium	Wheat bran	[28]
Bacillus sp. strain BK	Bacterium	Oat spelt xylan	[29]
Trichoderma viride	Fungus	Bajra straw, wheat straw, maize straw, barseem hay, oat hay and jowar straw	[30]
Aspergillus niger AN-13	Fungus	Wheat bran	[31]
Streptomyces sp. (strain Ib 24D)	Bacterium	Tomato pomace	[32]
Trichoderma spp. FETL c3-2	Fungus	sugar cane baggase, palm kernel cake	[33]
Thermoascus aurantiacus	Fungus	Wheat bran, sugarcane bagasse, orange bagasse, corncob, green grass, dried grass, sawdust and corn straw	[34]

Goyal et al. [30] isolated a *Trichoderma viride* culture and different agroindustrial residues such as bajra straw, wheat straw, maize straw, barseem hay, oat hay and jowar straw were tested for the production of xylanase. Maize straw was proved to be the best inducer for the xylanase production. Sodium nitrate was the best nitrogen source among all inorganic (ammonium sulfate, ammonium nitrate, sodium nitrate and ammonium chloride) and organic (urea and peptone) sources. *Thermoascus aurantiacus* was used to produce cellulase and xylanase simultaneously using different agro-industrial residues (wheat bran, sugarcane bagasse, orange bagasse, corncob, green grass, dried grass, sawdust and corn straw) as sole substrate. Corn cob, grasses and corn straw (without enrichment) showed highest level of enzyme production, where more xylanolytic enzymes were generated than cellulolytic enzymes [34]. The enzymes were active in both acidic and basic pH and remained stable for 1 h at 60°C.

Applications of Xylanases

Xylanase preparation is useful in animal feed processing to increase the digestibility, in food processing for clarification, wine industry to increase the extraction, to increase the quality of bread, extraction of bioactive compounds and in textile and paper industries. Plant biomass and other agricultural byproducts can be hydrolyzed by xylanase enzymes to low molecular sugars which in turn serve as chemical and fuel feedstock. Some of the important reported applications of xylanase are given in Table 2.

Saccharification of Agroresidues

Utilization of agro-industrial residues reduces the cost of production of value added product such as biofuels [48]. Different agro-industrial residues can be effectively hydrolyzed using a mixture of hydrolyzing enzymes including xylanase. The product can be utilized for biotechnological production of value added products [43]. Pal et al. [43] reported that pretreatment of agroindustrial residues is a necessary factor to achieve maximum saccharification as this process allows removal of lignin, reduces cellulose crystallinity, and increases the porosity of substrate. Pal et al. [43] tested the production of xylanase using different agroresidues such as mustard stalk and straw, wheat bran and rice straw. *Termitomyces clypeatus* could grow and produce multi enzymes such as CMCase, β-glucosidase, xylanase and β-xylosidase. This enzyme preparation was used for the saccharification of different agroresidues.

Plant Biomass Hydrolysate for Biofuel Production

Saccharified plant biomass can be used for the production of ethanol. Ethanol can be used as a blend or as such for the fuel purposes. A truncated xylanase gene was expressed in ethanol producing organisms such as *Escherichia coli* and *Klebsiella oxytoca* and studied their effect on saccharification and ethanolproduction by two step process [49]. The expressed intracellular xylanase in the cell wall was collected by harvesting the cells and used for saccharification xylan at high temperature. After saccharification, the temperature was reduced to 30°C and the hydrolysate was fermented with same organisms. Genetically engineered *Klebsiella oxytoca* possess both systems for transport of monomer, dimer and trimer of xylose and fermentation of these sugars to ethanol [49]. Incomplete degradation of xylose, insufficient expression of enzymes, ethanol intolerance of organisms will affect the conversion of agroindustrial residues to ethanol.

Table 2. List of major industrial applications of xylanases

Industry/Application sector	Action/Advantage	References
Pigment extraction	Tissue degradation of root of *Lithospermum erythrorhizon* increased the extraction of natural red dyes. The enzyme treatment reduced the batch extraction cost.	[40]
Wine making industry	Improved extraction (Mixture of cellulose, xylanase and pectinase)	[41]
Animal feed	Improved digestibility by reduce the viscosity	http://www.megazyme.com/downloads/en/data/K-XYLS.pdf
Pig feed	Improved digestibility	[42]
Biomass treatment/hydrolysis	Mustard plant waste saccharification; Improved sugar yield on pretreated stalk and straw	[43]
Pulp and paper industry	biobleaching effect on wheat straw pulp	[12]
Food industry	Improvement the texture of bread	[44]
Biomass treatment/hydrolysis	Liberated mainly xylotriose from birchwood xylan and wheat bran; have ability to hydrolyse, xylobiose, xylotriose, xylotetraose, xylopentaose, and xylohexaose	[8]
Biomass treatment/hydrolysis	Production of feruloyl oligosaccharides from wheat bran insoluble dietary fibre by xylanase from *Bacillus subtilis*	[45]
Biomass treatment/hydrolysis	Agricultural residues such as corn hull, sugarcane bagasse, rice straw, corn cop, rice husk and rice bran; unbleached sugarcane bagasse, eucalyptus and pine kraft pulps hydrolysed with *Bacillus* sp. strain BK produced extracellular cellulase-free xylanolytic enzymes	[29]
Biomass treatment/hydrolysis	Bajra straw, wheat straw, maize straw, barseem hay, oat hay and jowar straw were susceptible to the xylanase enzyme produced by *T. viride*	[30]
Fruit processing	Clarification of fruit juice	[46]
Biomass treatment/hydrolysis	Wood pulp industry waste was used to make hydrolysate for the growth of ethanol producing organisms and spent fibre was used as a substrate for the production of xylanase	[47]

Cavka et al. [47] developed a refinery concept for the production of ethanol biofuel and hydrolyzing enzymes from lignocellulosic sludge and waste fibre from pulp industry. Hydrolysed sludge was used as a carbon source for the production of ethanol by *Saccharomyces cerevisiae*. The spent fiber sludge hydrolysate was used as a medium for the growth of recombinant *Aspergillus niger* expressing the xylanase-encoding gene.

Shin et al. [50] used binary culture of *E. coli* for the production of ethanol from hemicellulosic residues. Co-cultivation of the two strains converted xylan hemicellulose to ethanol with a yield about 55% of the theoretical value (first strain with genetically engineered for expressing two hemicellulases and second one utilize xylooligosaccharide for ethanol production). Inclusion of three more hemicellulases improved the ethanol yield to 70%. Lee et al. [51] isolated different microorganisms for xylanolytic activity and ethanol production. Two *Pichia stipitis* strains could produce direct ethanol from xylan of about 60% of theoretical value.

Xylanases in Animal Feed

Xylanases are used in animal feed processing to increase digestibility. It is used in animal feed along with other hydrolyzing enzymes, such as glucanases, pectinases, cellulases, proteases, amylases, phytase, galactosidases and lipases. These enzymes synergistically act on complex biomass and reduce its viscosity by digesting the polymeric compounds. Arabinoxylan is one of the major content in most cereal solutions and slurries of having high viscosity. It reduces the rate of nutrient absorption in animals (http://www.megazyme.com/downloads/en/data/K-XYLS.pdf). Besides this, it increases the absorption of micro and macro element absorption and reduces this content in excreta and thus reduces the environmental pollution. One report shows the reduction of pathogenic infection in poultry on using feed treated with xylanase. Xylanase and essential oil supplementation reduced the incidence of horizontal transmission of *Salmonella* bacteria infection between chickens [52].

Forages and agroresidues can be treated with cellulase and hemicellulase for efficient utilization of agrobiomass for the absorption of nutrients and reduction of these factors in dung. Thus reduce the cost of feed and save environment in an efficient way.

Food Industrial Applications

Xylanase are used in food industry to improve dough qualities, increase digestibility, extraction of pigments and other bioactive compounds, juice and wine clarification, improve fermentation processes etc. They improve dough characteristics and bread quality leading to improved bread volume, crumb structure and reduce stickiness [53]. Xylanase treated bread has improved shelf life. Xylanase act on the water insoluble hemicellulosic part of the dough and make them soluble. They increase the water holding capacity and the dough becomes finer and forms more uniform crumbs. Xylanase from *Aspergillus* sp. improved bread qualities, such as finer texture of the crumb, extending freshness of bread, improving the colour and flavour and by improving the slicing ability [54].

Xylanase is used by wine industry for clarification and increasing the extractability. Commercial enzymes can digest the plant cell wall materials and improve the quality of juice and wine [41]. Byarugaba-Bazirake, [41] used recombinant yeast strains produce pectinase, xylanase, glucanase and amylase for juice and wine extraction and product clarification. Fungal xylanase gene was expressed in wine yeast strain and the toxicity was examined the toxicity in rats. According to Picó et al. [55] no adverse effects were detected by oral administration of xylanase from *Aspergillus nidulans* expressed in a wine yeast strain.

Xylanase from *Bacillus stearothermophilus* was used for the clarification of citrus fruit juice. There was increase in sugar release and decrease in turbidity by 52.97% and 35.34% respectively [56].

Xylanase is used to improve the extractability of pigments and bioactive compounds from biomass. There is a need of repeated extraction of these compounds for getting maximum yield while using usual extraction procedure. The disruption of epidermal matrix of root of *Lithospermum erythrorhizon* increased the extractability of red pigment. There was no difference in color quality when enzyme treated extraction and traditional extraction [40]. Extraction of essential oils, aroma etc. can be improved by xylanase treatment.

Applications in Chemical Industries

Xylose, the monomeric sugar from xylan, can be converted into other chemical molecules as ethanol. The other major chemical compounds are lactic acid, acetic acid, butanediol, xylitol etc. 2,3-Butanediol is a precursor for biofuel and it acts as a good solvent. Dehydration of 2,3-butanediol results in the formation of methyl ethyl ketone and further dehydration results in 1,3-butadiene and which is precursor of many polymeric substances [57].

Lactic acid production is carried out usually with six carbon sugars especially glucose from starch [58-61]. There are also certain studies where use of lignocellulosic biomass for lactic acid production [61-64]. Lactic acid is a chemical having versatile applications in food, pharmaceutical, textile, leather, and other chemical industries. Garde et al. 2002 [65], evaluated the wheat straw hemicellulose hydrolysate for the production of lactic acid using *Lactobacillus brevis* and *Lactobacillus pentosus*. The coinoculation improved lactic acid yield as the spectrum of utilization of carbon sources is different in different strains. Walton et al. [66] used *Bacillus coagulans* MXL-9, a spore forming bacteria, for lactic acid fermentation using pre-pulping hemicellulose extracts. This bacterium had high tolerance against inhibitors such as acetic acid and sodium, which are present in pre-pulping hemicelluloses extracts and could consume pentose and hexose sugars.

Enzymatically (using endo and exoxylanases) hydrolysed steam exploded populus plant hemicellulosic fraction was used for the production of propionic acid by *Propionibacterium acidipropionici*. The propionic acid yield was 18 g L- 1 from 29.6 gL^{-1} xylose equivalents [67]. Brownell & Nakas, [68] used *Clostridium thermoaceticum* to ferment hydrolysate of pretreated oat splet xylan and hemicellulose isolated from hybrid poplar. 42gL^{-1} acetic acid was produced after 116h from approximately 2% xylose at pH 7.

Xylitol is a low calorie sweetener recommended for diabetic patients. It has teeth hardening property and antimicrobial activity, so it is used in orthodontic applications [57]. Canilha et al. [69], produced xylitol from wheat straw hemicelluloses hydrolysate and Alves et al.,[70] pretreated sugarcane bagasse hemicellulosic hydrolysate for xylitol production using *Candida guilliermondii*. A process have been developed for the continuous production of xylitol from hemicellulosic hydrolysate utilizing *C. guilliermondii* cells immobilized onto natural sugar cane bagasse fibres [71]. Carvalho et al. [72] developed immobilization technique by *C. guilliermondii* cells entrapped in Ca-alginate for the use of sugarcane bagasse hydrolysate as a raw material for xylitol production.

Textile Industry Application

The xylanolytic enzymes can be used in the textile industry to process natural fibers, such as jute. Cellulase free xylanase is used in this purpose. Xylanase treatment helps increase the water absorbing properties of fibre and increase the removal rate of impurities present in the wall material. The treatment reduces the use of bleaching agent. According to Dhiman et al. [73], enzymatic treatment increase smoothness and softness of fibre after desizing and bioscouring without affecting the strength of fibres.

Wood Pulp and Paper Industrial Applications

Xylanase helps in selective degradation of hemicelluloses and of wood parts and increases the extractability of lignin and improve the whiteness of the pulp. The bleaching process is faster while the pulp is treated with xylanase. Use of xylanase in pulp treatment help to reduce the use of environmentally hazardous bleaching agent like chlorine dioxide and other bleaching agent and thus reduce the cost of chemicals [74]. To preserve the cellulose strength of the pulp, the enzyme preparation must be free from the contamination from cellulase. Cellulase along with xylanase can be used for the deinking of the used paper. But cellulase use will negatively influence the paper quality and increase the COD of the effluent.

Conclusion

Xylanase is a versatile enzyme with application in many of the industries with increasing interest in the level of applications. Together with other enzymes, xylanase can be used to increase the quality of bread, liquefaction of fruit and vegetables, clarification of juices, biobleaching, saccharification of agroindustrial residues etc. Biological production and application of these enzymes by using agroindustrial residues open new avenue for the production of value-added products and reduction in the environmental pollution.

There is an exploding population for the demand of commodity chemicals and fuels and generation of agroindustrial residue increase day by day. Thus, to maintain safe environment, these residues can be treated in an eco-friendly way and utilized for the production of chemical feedstock and biofuels, such as ethanol. There are different unexploited or poorly exploited wastes, such asmango peel, tender coconut waste, dried leaves waste, banana peel etc. We have to adopt zero waste policy to exploit all these agricultural or agroindustrial waste for value added products production with the use of enzymes such as xylanase or organisms producing xylanase.

Enzymes are needed in bulk quantity for the environmental applications and for hydrolysis of hemicelluloses. Hence, the production must be inexpensive for the large scale applications. Enzyme cocktail qualities, such as presence or absence of cellulase, pH and temperature optima etc. has to be evaluated before application. For the food and feed industrial uses GRAS status must be maintained and the effect of these in target organisms must be evaluated. Metabolic or genetic engineering will favour the generation of recombinant strains with desirable characteristics. Single organism can perform the direct conversion of hemicellulose to product, and for this an organism need many enzymes. The efficient organism can be mutated directly for the deficient genes to solve this problem or co-cultivation of the superior strains can be done for effective utilization. The improvement in the biocatalysis can be done with knowledge of relation between structure and function of enzymes. There is a need for exploitation of these enzymes for novel applications. The disposal or management of waste generated from xylanase production and saccharification processes are again a hurdle for this endeavor. The wastes generated from the xylanase production can be again utilized for saccharification processes by other enzymes with respect to the nature of residual molecules. These wastes can be converted to fertilizers, such as compost and utilized for the agricultural purposes.

REFERENCES

[1] Khucharoenphaisan, K (2009) . Purification, characterization and induction of xylanase from thermomyces lanuginosus isolated in Thailand. PhD Thesis. Kasetsart University. http://research.rdi.ku.ac.th/world/cache/11/KhwanchaiKHUAll.pdf

[2] Collins, T., Gerday, C. & Feller, G. *FEMS Microbiology Reviews*, 29, 3 (2005).

[3] Johannson, M.H. & Samuelson O. *Wood Science and Technology,* 11, 251 (1977).

[4] Puls, J. & Schuseil J. Chemistry of hemicelluloses: relationship between hemicellulose structure and enzymes required for hydrolysis. *In*: Hemicellulose and hemicellulase. M.P. Coughlan and G. P. Hazlewood (Eds.). Portland Press, London. pp 1-28. (1993).

[5] Kar, S. Mandal, A. das Mohapatra, P. K., Mondal K. C. & Pati B. R. *Brazilian Journal of Microbiology* 37, 462 (2006).

[6] Azeri, C., Tamer A. U. & Oskay M. *African Journal of Biotechnology* 9, 63 (2010).

[7] Kavya, V. & Padmavathi, T. *Polish Journal of Microbiology* 58, 125 (2009).

[8] Dai X.-J., Liu M.-Q., Jin H.-X., Jing M.-Y. *Czech Journal of Food Science,* 29,557 (2011).

[9] Kaushal, R., Sharma, N. & Tandon, D. *Turkish Journal of Biochemistry*, 37, 35–41. (2012).

[10] Subramaniyan S. & Prema P. *Critical reviews in Biotechnology,* 22, 33-64. (2002).

[11] Gomes J., Purkarthofer H., Hayn M, Kapplmuller J., Sinner M. & Steiner W. *Applied Microbiology Biotechnology,* 39, 700 (1993).

[12] Li, X.T., Jiang, Z.Q., Li, L.T., Yang, S.Q., Feng, W.Y., Fan, J.Y. & Kusakabe, I. (2005) *Bioresource Technology,* 96, 1370-1379.

[13] Khucharoenphaisan, K., Tokuyama, S. & Kitpreechavanich, V. Characterization of the thermostability of xylanase produced by new isolates of *Thermomyces lanuginosus. ScienceAsia* 34, 187 (2008).

[14] Sun, Y. & Cheng, J. *Bioresource Technology*, 83, 1 (2002).

[15] Winterhalter, C. & Liebl, W. *Applied and Environmental Microbiology,* 61, 1810 (1995).

[16] Sharma, S., Adhikari, S. & Satyanarayana, T. *World Journal of Microbiology Biotechnology,* 23, 483. (2007).

[17] Zhang, S., Zhang, K., Chen, X., Chu, X., Sun, F. & Dong, Z. *Biochemistry Biophysics Research Communications*, 395, 200 (2010a).

[18] Zhang, Z.G., Yi, Z.L., Pei, X.Q. & Wu, Z.L. *Bioresource Technology,* 101, 9272(2010b).

[19] Subramaniyan S. Studies on the production of bacterial xylanases. PhD Thesis, Cochin University of Science and Technology, Kerala, India. (2000).

[20] Smith, D.C. & Wood, T.M. *Biotechnology Bioengineering*, 38, 883 (1991).

[21] Beg, Q.A., Kapoor, M., Mahajan, G. & Hoondal, S. *Applied Microbiology Biotechnol.* 56, 326 (2001).

[22] Ratto, M., Poutanen, K & Viikari, L. *Applied Microbiology and Biotechnology* 37, 470 (1992).

[23] Copa-Patino, J.L., Kim, Y.G. & Broda, P. *Applied Microbiology and Biotechnology,* 40, 69 (1993).

[24] Steiner, W., Lafferty, RM, Gomes, I., & Esterbauer H. *Biotechnology and Bioengineering,* 30, 169 (1987).

[25] Bakri, Y, Al-Jazairi M & Al-Kayat G. *Polish Journal of Microbiology* 57, 249 (2008).

[26] Mountfort D and Asher R. *Applied and Environmental Microbiology*, 55, 1016 (1989)

[27] Assamoi A.A., Delvigne F., Aldric J-M, Destain J. & P. Thonart. *Biotechnol. Agron. Soc. Environ.* 12, 111 (2008).

[28] Bajaj, B.K., Yash Paul Khajuria & Vijay P. Singh., (2012). *Biocatal. Agric. Biotechnol.,* http://dx.doi.org/10.1016/j.bcab.2012.05.001

[29] Kaewintajuk K., J. Kongkiattikajorn,_Ratanakhanokchai K. and Kyu K. L. 2005. Hydrolysis of Agricultural Residues and Kraft Pulps by Xylanolytic Enzymes from Alkaliphilic *Bacillu*s sp. Strain BK. 1st international conference on Fermentation Technology for Value Added Agricultural Product. www.202.28.96.69/ inter/images/Fervaap2005/Oral-Poster-Full/fullpaper/oral/NF/O-NF39FP.pdf

[30] Goyal M.; K.L. Kalra; V.K. Sareen; G. Soni. *Brazilian Journal of Microbiology* 39,535 (2008).

[31] Cao, Y., D-J Meng, Lu J. & Long, J. *African Journal of Biotechnology* 7, 631 (2008).

[32] Rawashdeh R., I. Saadoun and A. Mahasneh. *African Journal of Biotechnology* 4, 251 (2005).

[33] Pang, P. K., Darah. I., Pop pe, L., Szakacs, G. & Ibrahim, C.O. *Malaysian Journal of Microbiology*, 2, 7 (2006).

[34] da Silva R. ; Lago E. S.; Merheb C. W.; Macchione M. M.; Park Y. K.& E. Gomes *Brazilian Journal of Microbiology* 36, 235 (2005).

[35] Archana, A. & Satyanarayana, T. *Enzyme and Microbial Technol*ogy, 21, 12 (1997).

[36] Yasser B., Philippe J. & Thonart P. *Applied Biochemistry and Biotechnology*, 108, (2003).

[37] Priem, B., Dobberstein J. & Emeis, C.C. *Biotechnology Letters*, 13, 149(1991).

[38] Yang, V.W., Zhuang, Z., Elegir, G. & Jeffries, T.W. *Journal of Industrial Microbiology* 15, 434 (1995).

[39] Sudan, R. & Bajaj, B.K., *World Journal of Microbiology Biotechnology*, 23, 491 (2007).

[40] Kim D. H., Kim J. H., Bae S. E., Seo J. H., Oh T.K., & C. H. Lee. *J. Agric. Food Chem.*, 53, 2541(2005).

[41] Byarugaba-Bazirake, 2008 . The effect of enzymatic processing on banana juice and wine. Doctoral thesis. University of Stellenbosch.

[42] Kiarie E., Owusu-Asiedu A., Peron A., Simmins P.H., Nyachoti C.M. *Livestock Science,* 148, 129. (2012).

[43] Pal S., Banik S. P. & Khowala S. *Industrial Crops and Products* 41, 283 (2013).

[44] Jiang, Z.Q., Yang, S.Q., Tan, S.S., Li, L.T., Li, X.T. *Letters in Applied Microbiology,* 41, 69 (2005).

[45] Yuan X.P., Wang J. & Yao H.Y. *Food Chemistry*, 95, 484 (2006).

[46] Hang, Y.D. & Woodams, E. E. *Letters in Applied Microbiology,* 24, 389 (1997).

[47] Cavka A, Alriksson B, Rose SH, van Zyl WH & Jönsson LJ. *Journal of Industrial Microbiology and Biotechnology*. 38, 891 (2011).

[48] van Wyk, J.P.H. *Trends in Biotechnology* 19, 172 (2001).

[49] Burchhardt G. & Ingram LO *Applied and Environmental Microbiology*, 58, 1128 (1992)

[50] Shin H-D, McClendon S., T. Vo, & R. R. Chen. *Applied and Environmental Microbiology*, 76, 8150 (2010).

[51] Lee H., Biely P., R. K. Latta, Maria F. S. Barbosa & Henry Schneider. *Applied and Environmental Microbiology,* 52, 320 (1986).

[52] Amerah AM, Mathis G & Hofacre CL. *Poultry Science*, 91, 943 (2012).

[53] Butt, M.S., M. Tahir-Nadeem & Z. Ahmad. *Food Technology Biotechnology,* 46, 22 (2008).

[54] Chereji R., Căpriţă R. & I. Creţescu *Animal Science and Biotechnologies*, 43, 379 (2010)

[55] Picó Y, Fernández M, Rodríguez R, Almudéver J, Mañes J, Font G, Marín R, Carda C, Manzanares P & Ramón D. *J Agric Food Chem*. 47, 1597 (1999).

[56] Dhiman S. S., Garg G., J. Sharma & R. Mahajan. *New Biotechnology,* 28, 746 (2011).

[57] Marais, S. (2008). Enzymatic hydrolysis with commercial enzymes of a xylan extracted from hardwood pulp. Master's thesis. University of Pretoria, Pretoria.

[58] John R.P., Nampoothiri, K.M. & Pandey, A. *Proc Biochem* 41,759 (2006a).

[59] John R.P., Nampoothiri, K.M. & Pandey, A. *Appl Biochem Biotechnol* 134, 263 (2006b).

[60] John R.P., Nampoothiri, K.M. & Pandey, A. *Biotechnol Lett* 28, 1823 (2006c).

[61] John R.P., Nampoothiri, K.M. & Pandey, A. *Appl Microbiol Biotechnol*.74, 524(2007).

[62] Miura S, Arimura T, Itoda N, Dwiarti L, Feng JB, Bin CH & Okabe M. *J Biosci Bioeng*, 97, 153 (2004)

[63] Moldes AB, Alonso JL & Parajo JC. *Bioproc Eng* 22,175 (2000).

[64] Moritz J.W. & Duff S.J.B. *Biotechnol Bioeng*, 49, 504 (1996).

[65] Garde, A., Jonsson, G., Schmidt A.S., & Ahring, B. K. *Bioresource Technology*, 81, 217 (2002).

[66] Walton S. L. , Bischoff K.M. , van Heiningen A. R. P. & van Walsum G. P. *Journal of Industrial Microbiologyl Biotechnology* 37, 823 (2010).

[67] Ramsay J. A., Aly Hassan M.-C., & Ramsay B. A. *Enzyme and Microbial Technology* 22, 292 (1998).
[68] Brownell J. E. & Nakas J. P. *Journal of Industrial Microbiology & Biotechnology*. 7, 1-6. (1991).
[69] Canilha L, Carvalho W, Almeida E & Silva JB *Journal of Science of Food and Agriculture,* 86, 1371 (2006).
[70] Alves LA, Felipe MGA, Silva JBAE, Silva SS & Prata AMR *Appl Biochem Biotechnol* 70–72, 89 (1998).
[71] Diego T.S., Boutros F.S., Juan D.R., Attilio, C. & Silvio, S.S. *Journal of Food Engineering,* 86, 542 (2008).
[72] Carvalho W, Santos JC, Canilha L, Silva SS, Perego P, Converti A. *Biochemical Engineering Journal* 25, 25 (2005).
[73] Dhiman, S. S., J. Sharma & Buttan B. *BioResources* 3, 1377 (2008).
[74] Tolan J.S. & Collins J. *Pulp & Paper Canada*, 105, 44 (2004).

In: Enzymes in Value-Addition of Wastes
Editors: Satinder Kaur Brar and Mausam Verma
ISBN: 978-1-62948-968-1

Chapter 9

ENZYMATIC PRODUCTION OF VALUE-ADDED PRODUCTS FROM CASHEW APPLE JUICE

Khasim Beebi Shaik [*] ***and Uma Talasila***
Department of Biotechnology, GITAM Institute of Technology, GITAM University, Gandhinagar, Visakhapatnam, Andhra Pradesh, India

ABSTRACT

The average annual production of cashew apples (*Anacardium occidentale*) in the world is estimated to be 30 million metric tonnes. Cashew apples yield about 78 % (w/v) juice and are rich in carbohydrates (10-12%), minerals (261 mg/100 g) and vitamin C. Despite being highly nutritive, poor shelf life and astringency of the fruits limits their utilization and hence are wasted in the fields. Owing to the availability of cashew apples in abundance, together with its rich composition (reducing sugar, fibers, vitamins and mineral salts) turns cashew apple juice into an inexpensive culture medium to produce different value added products. Many studies were conducted on microbial fermentation of cashew apple juice for the production of ethanol and our study produced a yield of 7.8 % (w/v) ethanol. However, studies pertaining to the use of enzymes for the production of ethanol from cashew apple juice remain unexploited. Advances in the field of enzyme technology have opened up new horizons for the production of different products with novel applications.

The ability of yeast cells to convert sugars present in the cashew apple juice into ethanol are mediated via the enzymes namely, invertase and zymase. Enzymatic production of ethanol might improve the yield and productivity of the desired product. In this direction, the present chapter is an attempt to utilize enzymes for the development of value-added products from cashew apple juice to meet different end uses, which in turn, would not only minimize the wastage of cashew apples, but would also generate an additional income to the cashew farmers. The techno-economic feasibility of the production of these products along with the advantages and disadvantages of the conversion processes are described in detail.

[*] Corresponding author: +91 9347973429 (Mobile); E-mail ID: khasimb@gitam.edu.

Keywords: Cashew apple juice, enzymatic conversion, ethanol, value-addition, economics

INTRODUCTION

Cashew is cultivated in 32 countries of the world with Brazil, India, Vietnam and Nigeria as the main producers (Honorato and Rodrigues, 2008). The cashew tree grows even on poor soils with low rainfall. Approximately 1 ton/hectare of raw cashew nut and 10 ton/hectare cashew apples is the expected yield of cashew tree under rainy conditions (Silveira et al., 2010).

While it is estimated that about 30 million metric tons of cashew apples are produced yearly in the world and 20 lakh tons in India, 85 % of the crop goes to waste (Michodjehoun-Mestres et al. 2009, Mandal et al., 1985). Generally, the cashew apples with nuts attached fall from the trees when ripe and are gathered from the ground, the nuts are separated and the cashew apples are discarded. About 90 % of cashew apple production is being wasted in all parts of the world. In Brazil, more than 80 % of its production is wasted. The present consumption of cashew apples either in raw or processed form is about 10 % of its production (Oduwole et al., 2001).

Cashew apple is a non-climacteric fruit found in three colors: yellow, orange and red, with the same pale yellow pulp, weighs about 75-80 g and is 6-10 cm in length (Maciel et al. 1986). It is fibrous, juicy and highly astringent. The edible cashew "apple" is the thick swollen receptacle/hypocarp or peduncle, to which the kidney-shaped cashew "nut" is attached. These are highly nutritious and possess potent therapeutic value. However, cashew apple is considered an agricultural byproduct of cashew nut industry.

Neither the cashew apples nor the juice extracted from the fruits is utilized, despite possessing nutritional properties thereby; large amounts of cashew apples are being wasted in the field after nut separation. Hence, the present chapter attempts to review the nutritional quality of cashew apple juice and the factors for spoilage of cashew apple juice. The chapter also deals with the available literature for its utilization as fermentation medium for the production of value added products suitable to meet different end uses in various sectors. Further attempts have been made to study the feasibility of use of enzymes for conversion of sugars present in the juice for the development of such products in order to replace the traditional microbial fermentation processes.

QUALITIES RESPONSIBLE FOR THE PRODUCTION OF VALUE ADDED PRODUCTS

The average yield of juice extraction is approximately 85 % (Silveira et al., 2010). Each cashew apple yields about 25 ml of juice composed of good amount of nutrients. Cashew apples are considered as a good source of reducing sugars (glucose and fructose), minerals, amino acids, carotenoids, phenolics (quercetin, anacardic acid, tannins), organic acids, vitamins particularly vitamin C and anti-oxidants (Azevedo and Rodrigues 2000; da Silva et al. 2000; Campos et al. 2002; Azevedo and Rodrigues 2005; Cavalcante et al. 2005; Trevisan et al. 2006; de Carvalho et al. 2007; Honorato and Rodrigues 2008). Owing to the availability of cashew apples in abundance, together with its rich composition (reducing sugars, fibers, vitamins, and mineral salts), turns cashew apple juice into an inexpensive culture medium to produce value added products by economical fermentation methods and can be considered as a substitute in providing the microbial mineral requirements (Rocha *et al.*, 2006). The chemical composition of the juice responsible for its action as fermentation medium is shown in Table 1.

The vitamin C content of cashew apples is about six times than that of an orange (da Silva et al. 2000; Azam Ali and Judge 2001; Azoubel et al. 2005). During maturation of cashew apple, maximum ascorbic acid content was observed in 55-days fruit, which can be attributed to the decrease in the ascorbic acid oxidase activity (Chempakam 1983). Elongated red and yellow fruits collected from the Northeast and the Southeast regions of Brazil presented slightly higher vitamin C content (114 mg/100 g) than the rounded red ones (104 mg/100 g) (Assuncao and Mercandante 2003). In addition to vitamin C, cashew apple juice contains thiamine, niacin, riboflavin and precursors of vitamin A. Cashew apple juice is also found to be good source of minerals, such as copper, zinc, sodium, potassium, calcium, iron, phosphorous and magnesium (Lowor and Agyente-Badu 2009). In addition to these minerals, the juice also contains sulphur, silicon, chlorine, aluminium and bromine (Marc et al. 2011).

Table 1. Chemical composition of cashew apple juice

Composition	Constituents	Range	Units	Reference
pH		3.67 - 4.53	-	Gunjate and Patwardhan (1995); Osho (1995); Oduwole et al. (2001); Campos et al. (2002); Azoubel et al. (2005); Lowor and Agyente-Badu (2009)
TSS		7.4 - 12.8	°Brix	Gunjate and Patwardhan (1995); Oduwole et al. (2001); Campos et al. (2002)
Moisture		90.4%	g/100 g	Azoubel et al. (2005)
Ash		0.19 – 0.45	g/100 g	Osho (1995)
Specific gravity		1.030	-	Osho (1995)
Refractive index		1.339	-	Osho (1995)
Sugars				
Total sugars (sucrose, maltose, raffinose and two higher saccharides (unidentified)		6.3 – 9.9	g/100 g	Azoubel et al. (2005); Damasceno et al. (2008)
Reducing sugars (glucose and fructose)		6.24 – 9.8	g/100 g	Azoubel et al. (2005); Damasceno et al. (2008)
Vitamins				
Vitamin C		126 – 372	mg/100 ml	Price et al. (1975); Morton and dowling (1987); Gunjate and patwardhan (1995); Damasceno et al. (2008)
Minerals				
Ca, P, Fe		0.9 – 21.4	mg/100 g	Morton and Dowling (1987)
K		1.53	g/L	Osho (1995)
Mg, Zn, Na		16-105	g/L	

SPOILAGE OF CASHEW APPLES AND ITS JUICE

Post-harvest shelf-life of cashew apples is very short as a result neither the fruits nor the extracted juice could be stored, even at refrigerated conditions. The fruits are to be processed within two to three hours of procurement. The quality of the juice can be adversely affected by physical, chemical, biochemical and microbiological processes, resulting in astringency, sedimentation, browning and perishability, leading to wastage of the nutritious juice. Among these properties, perishability is considered as the main factor responsible for the spoilage of these fruits. Other constraints hindering the full utilization of the fruits are poor storability and lack of information on an appropriate preservation technology. Preservation of cashew apples and the juice seems to be very difficult in terms of stabilizing nutrients as well as maintaining microbial quality. Hence, there is a need to utilize the cashew apple waste as fermentation medium for the development of value added products. In order to prevent the spoilage of the nutritious fruits, the juice is utilized for the preparation of various products.

PERISHABILITY

Besides possessing various antimicrobial compounds, such as anacardic acids and volatile compounds, cashew apples are highly perishable, cannot be transported over long distances and need to be processed at the farm settlements (Akinwale et al. 2001). Microbial spoilage can be characterized either by their growth and/or by their metabolic processes. As a consequence of microbiological changes, the quality of the juice is adversely affected. The microbial spoilage of fruits can be due to the presence/invasion of bacteria, yeasts and molds. The presence of yeasts and moulds namely, *Rhodotorula sp., Cryptococcus sp., Torulopsis sp., Candida sp., Saccharomyces sp, Mucor sp., Aspergillus niger, Rhizopus sp.* and *Colletotrichum gloeosporoides* enhances the spoilage of cashew apples (Bopaiah 1985). Formation of organic acids and generation of volatile compounds, such as ethanol from fermentation of sugars are mechanisms that result in aroma and flavor change. The minute concentrations of antifungal compounds, such as (*E*)-2-hexenal, furfural, benzaldehyde, and 2-methylpentan-1-ol in the cashew apple are not potent enough for defence against fungal action (Muroi et al. 1993). Sterile filtration was however, found to be efficient in removing microbial contamination from the juice (Talasila et al. 2011).

UTILIZATION OF CASHEW APPLES AND ITS JUICE IN VARIOUS SECTORS

The main form of utilization of cashew apples is to process them into juice, but its high astringency seems to limit its acceptance. The astringent taste in cashew apple can be attributed to tannins and other substances, which are found in the waxy layer of the skin (Bidaisee and Badrie, 2001). Various products developed from cashew apples and presently utilized in various sectors are discussed in detail. Scientists have standardized methods for the preparation of various products: cashew apple syrup, squash, jam, candy, chutney, pickles and juice (Augustin 1982, Rale 1985). Natural/fresh cashew apple juice and cashew apple products, such as cashew apple ice creams, concentrated juice, frozen pulp, nectar, ready-to-drink and cashew apple juice by the name 'Cajuina' and other foodstuffs are popular in Brazil (Damasceno et al., 2008; Fontes et al., 2009). Some of the cashew apple juice products are shown in Figure 1. In tropical America, cashew apple juice is consumed directly or is used in ices and jellies. In Venezuela, cashew apples are sliced, stewed or made into fruit soup. Cashew wine is bottled commercially in Cuba, Costa Rica, Brazil, the Philippine Islands and India. Efforts have been made by Central Food Technological Research Institute, Mysore, India to develop processes for converting cashew apple into beverages (Morton 1961).

Source: Google.com.

Figure 1. Commercial products of cashew apple juice.

Cashew Apple Juice as Fermentation Medium

Food Industry

Honorato et al., 2008 showed that cashew apple juice was used as medium for the growth of *Leuconostoc mesenteroides* B-512F for the production of dextran, lactic acid, mannitol and oligosaccharides. Maximum mannitol productivity (1.8 g/L/h) was obtained in cashew apple juice containing 50 g/L total reducing sugars after 10 hours of fermentation. *Lactobacillus casei* B442 was used for the growth (biomass) and production of lactic acid using cashew apple juice. 2.36 g/L/h lactic acid was produced under following conditions: 50 g/L reducing sugars, 6 g/L ammonium sulfate, pH 6.5 and 37 °C (Silveira et al., 2010). Osho (1995) observed the ability of cashew apple juice to support the growth of yeasts and suggested it as a feedstock for the production of single cell protein and wine. Cashew apple juice was used as a feedstock for the production of single cell protein using *Saccharomyces cerevisiae* NCYC 125 (Layokun et al., 1986).

Fermented cashew apple juice medium containing fructose, glucose, added sucrose, yeast extract and K_2HPO_4, comprising oligosaccharides was reported to possess prebiotic effect, as evaluated through the growth of *Lactobacillus johnsonii* B-2178. The development of such non-dairy prebiotic product can replace the dairy-based prebiotic products in the market and can be consumed by people who are allergic to milk and milk products (Vergara et al., 2010). Pereira et al. (2010) revealed that fermented cashew apple juice with *Lactobacillus casei* NRRL B-442 is an alternative functional food containing probiotics. At pH 6.4 and temperature 30 °C, 7.48 log CFU/ml inoculum of *L. casei* in 100 ml of juice maintained at 16 h produced prebiotic cashew apple juice. The prebiotic juice was stable upto 42 days stored at 4 °C. The viable cell counts of *L. casei* were found to be higher than 8.00 log CFU/ml throughout 42 days.

Enzymes

The nutritive cashew apple juice can be a suitable low-cost substrate for the production of dextransucrase. Chagas et al. (2007) studied the use of cashew apple juice as substrate to produce dextransucrase using *Leuconostoc mesenteroides* B512F. Dextransucrase in cashew apple juice was highly stable at 30 °C and was suitable for industrial application (Honorato and Rodrigues, 2008). The enzyme was stable at room temperature (30 °C) at a pH of 5.0 with a maximum enzyme activity (1.95±0.01 – 2.93±0.02), 3.5 times highly active than that found in the synthetic medium (0.05 ±0.02 UDS/10 μl). Another author, Rabelo et al. (2009) produced dextransucrase using *Leuconostoc citreum* B-742 in fermented cashew apple juice as substrate. The authors found an optimal yield of approximately 80% for sucrose concentration lower than 60 g/L and reducing sugar concentration higher than 100 g/L.

Industrial Applications

Cashew apple juice was utilized as a substrate for the production of emulsan (biosurfactant), using *Acinetobacter calcoaceticus* RAG-1 and the results were compared with a defined mineral medium. The authors reported that biosurfactant was produced after 15 h of fermentation, with an emulsifying activity value of 58.8% after 34 h, showing that the biosurfactant could be able to emulsify kerosene (Rocha et al., 2006). Further, Rocha et al. (2007) attempted to produce rhamnolipid (biosurfactant) by *Pseudomonas aeruginosa* ATCC 10145 and observed a maximum biomass concentration of 1 g/L when cashew apple juice supplemented with peptone was used. The higher biosurfactant production occurred in this medium after 48 h of cultivation. Surface tension was reduced by 41% after 24 hours of cultivation. Further, the emulsifying activity (71.79%) of the biosurfactant was found to be higher in soy oil than kerosene (16.5%).

Rocha et al. (2009) produced 3.5 mg/L surfactin (biosurfactant) from *Bacillus subtilis* LAMI008 in mineral medium containing clarified cashew apple juice, supplemented with yeast extract after 24 h. The surfactin was capable of emulsifying kerosene (emulsification index 65 %) after 72 hours and the surface tension reduced from 21.37% after 24 h of cultivation at 30 °C. Giro et al. (2009) produced surfactin using clarified cashew apple juice supplemented with mineral medium in a conventional batch bioreactor using *Bacillus subtilis* LAMI008. The surfactin was produced after 48 h of fermentation, the surface tension (123 mg/L) reduced by 24.68 % at 72 hours and was able to emulsify kerosene, n-hexadecane and soybean oil, with 66.7 %, 64.52 % and 51.15 % emulsification index, respectively.

Alcoholic Beverage Industry

Cashew apple juice was used as a medium for the growth of microorganisms for the production of wine and vinegar. Akinwale (1999) used *Saccharomyces cerevisiae* and produced 10.6 % (v/v) alcohol, which was acceptable in terms of sensory evaluation (color, taste, flavor, overall acceptability) and chemical characteristics for six months. Muir-Beckford and Badrie (2000) prepared wine using *Saccharomyces cerevisiae* var. *ellipsoideus*. The wine was stored for seven days. The alcohol content of cashew dry wine (TSS: 4.10 ± 0.2 °Brix) was 11.59 ± 0.1 % whereas cashew sweet wine (TSS: 9.30 ± 0.2) was 11.86 ± 0.04 %. Microbial count of less than 10 CFU/ml was observed along with good consumer acceptance. Mohanty et al. (2006) used *Saccharomyces cerevisiae* var. *bayanus* to prepare cashew wine with 7 % alcohol, found to be acceptable as an alcoholic beverage. Fermentation was carried out at 32 ± 2 °C for six days and the wine was yellowish, acidic (1.21 g/100 ml as tartaric acid) and higher in tannin (1.9 ± 0.22 mg/100 ml). Silva et al. (2007) produced cashew wine (0.78 g/L/h) with 102.9 g/L alcohol concentration after 48 hours of fermentation in a stirred batch reactor. Further, response surface methodology was used for optimization of cashew wine vinegar fermentation and revealed 0.55 g/L/h wine with an ethanol concentration between 4.8–6 %. Attri (2009) standardized a technique for the production of wine from cashew apples by fermenting *Saccharomyces cerevisiae* var. *ellipsoideus* for 15 days. Araujo et al.(2010)developed wine using *Saccharomyces cerevisiae* SCP and SCT strains and carried out fermentation for six days. The percentage of alcohol produced with cashew apple juice inoculated with SCT strain was found to be 11 % (v/v) while SCP strain was 10 % (v/v). Joseph (2010) produced wine from preserved cashew apple juice (combination of addition of SMS 0.2 g/L and pasteurization at 90-95 °C for 1 min added to 3 months cold stored juice at -20 °C) using spontaneous and controlled fermentations using mixed microbial population and *Saccharomyces cerevisiae* var. *ellipsoideus* respectively for 3 months. Sensory evaluation reported that controlled wine was superior than spontaneous wine in terms of organoleptic and keeping qualities (with no microbial growth). The alcohol content of the juice in controlled fermentation was 8.3% whereas 7.08% in case of spontaneous fermentation.

Owing to high sugar content of cashew apple juice, Thiripurasundari and Usharani (2011) produced vinegar from cashew apple juice using immobilized cells of *Acetobacter aceti* MTCC 2975 via ethanol production using *Saccharomyces cerevisiae* (isolated from cashew apple waste). Different inert materials (bagasse, corn cobs and wood shavings) were compared for the immobilization (adsorption and entrapment) of *Acetobacter aceti* cells. Maximum fermentation with corn cobs was obtained after 28 days of fermentation and further, adsorbed cells showed better fermentation efficiency than entrapped cells.

Pharmaceutical Industry

Streptococcus zooepidemicus was inoculated in cashew apple juice for the production of hyaluronic acid. Cashew apple juice was compared with commercial Brain Heart Infusion (BHI) medium and observed that same cell mass was obtained when cashew apple juice was supplemented with yeast extract as that of BHI medium (Oliveira et al. 2005).

CASHEW APPLE JUICE AS CATTLE FEED

Cashew apple juice as well as effluent from cashew apple processing plants can be used as medium for the production of yeast biomass to enrich livestock feedstuffs (Rale, 1985). *Candida utilis* (NCIM 3336) produced more biomass (40 g/L) than *Saccharomyces cereviseae* (36 g/L) (occurring naturally on cashew apples).

Value-added Products

In order to avoid wastage of these fruits, low value, highly nutritious cashew apple juice was utilized for the production of different value-added products. In the present investigation, an attempt was made to utilize the sugars (6.2-9.9 g/100 g), amino acids (0.88–3.36 mM) and other micro nutrients such as minerals (Ca, P, Fe, Mg, Zn, Na, K) present in cashew apple juice by microbial fermentation into ethanol, single cell protein and hyaluronic acid. Feasibility of utilizing cashew apple juice as substrate for the production of value added products namely, ethanol, single cell protein and hyaluronic acid from different microorganisms was studied and the results obtained are presented as under. The study revealed that an optimum yield of 6.7 % ethanol, 1.115 g/L single cell protein and 0.028 g/L hyaluronic acid was obtained using *Saccharomyces cerevisiae* NCIM 3190, NCIM 3211 and *Bacillus subtilis* MTCC 441 respectively.

Ethanol

Table 2 represents the mean chemical composition of fresh cashew apple juice. The TSS content of the juice was found to be 13.7% Brix, which constitutes sugars (11.3%) and organic acids such as malic acid (0.318%). Owing to the sugar content of the juice, fermentation of cashew apple juice was carried out in the presence of *Saccharomyces cerevisiae* NCIM 3190 for the production of ethanol. The protein content of the juice was observed to be low (0.17%). The sugars and minerals present in the juice would be utilized for the production of desired products depending on the strain used through fermentation.

Table 2. Characteristics of cashew apple juice

Parameter	Value
pH	4.3
TTA (% malic acid)	0.318
TSS (% Brix)	13.7
Total sugars	11.3%
Protein	0.17%

Table 3 represents the optical density, sugars and ethanol content of fermented cashew apple juice. The total sugars of fresh cashew apple juice was found to be 11.6 g/L whereas the residual total sugars after 26 hours of fermentation was 2.4 g/L. The decrease in sugar content could be attributed to its utilization during fermentation.

Table 3. Optical density, sugars and ethanol content of fermented cashew apple juice

Fermentation time (hrs)	Parameter		
	O.D (610 nm)	Sugars (g/L)	Ethanol (%)
18	0.62	3.5	5.3
22	0.69	2.9	6.1
26	0.78	2.4	6.7
30	0.71	2.4	6.7

Sugars present in the juice alone were utilized as sole carbon source (without any nutrient supplementation) during fermentation. The total sugars present in fresh cashew apple juice were 11.6 g/L whereas residual sugar content after 26 hours of fermentation was found to be 2.4 g/L. Increase in the yeast cell count and the corresponding decrease in sugars was observed as the cells growing in logarithmic phase are metabolically active and reproduce at a faster rate. The increase in yeast cell count has increased the optical density of the medium from 0.49 to 0.78 upto 26 hours of fermentation. Increase in optical density with fermentation time could be clearly attributed to the microbial growth, as observed by the turbidity of the medium. Later, the cell count started decreasing, which is clearly shown from the optical density of the sample collected at the end of 30 hours.

All the collected samples were analyzed for percentage of ethanol using potassium dichromate method. Maximum ethanol concentration of 6.7 % (67 g/L) was obtained at 26 h of fermentation at pH 5.0 and temperature 28 °C. The same sample was preserved and analyzed using gas chromatography and the concentration was found to be 7.8 % (78.42 g/L).

The low yield of ethanol obtained from fermentation of cashew apple juice could be due to the toxicity of ethanol to *Saccharomyces cerevisiae*. Further studies are to be conducted to improve the yield of ethanol by supplementing minerals and their salts. Karuppaiya *et al.,* (2009) fermented cashew apple juice using *Zymomonas mobilis* under anaerobic conditions and obtained a maximum ethanol concentration of 12.64 g/L at pH 5.5, temperature 32 °C at 37 hours. The fermentation of cashew apple juice for 26 hours using immobilized *Saccharomyces cerevisiae* yielded an ethanol concentration of 7.64 % (Neelakandan and Usharani, 2009). Fermentation of cashew apple juice by *Saccharomyces cerevisiae* in an aerobic batch reactor at optimum conditions of pH 6.5, temperature 32 °C and 42 hours produced 15.64 g/L (Karuppaiya et al., 2010). On the other hand, a maximum ethanol yield of 0.488 g/g and productivity of 9.71 g/L/h was obtained using *Saccharomyces cerevisiae* at an initial sugar concentration of 87.7 g/L/h (Pinheiro et al., 2008). All the above results indicate the feasibility of using cashew apple juice as fermentation medium. Pacheco et al. (2010) fermented cashew apple juice using *Saccharomyces cerevisiae* immobilized in cashew apple bagasse. A yield of 5.49 g/L/h ethanol was produced with 95 % sugar consumption after 6 hours of fermentation. This high ethanol produced was until the 10th fermentation.

MICROBIAL FERMENTATION VS ENZYMATIC CONVERSION (USE OF ENZYMES FOR THE PRODUCTION OF VALUE ADDED PRODUCTS)

The process of breakdown of an organic compound into smaller molecules in the presence of yeast is known as fermentation. Fermentation of sugars (present in substrate) results in the product (for example, ethanol). The process of fermentation is initiated by yeast, which secretes two enzymes, invertase and zymase. These enzymes act as catalysts in the conversion process. Invertase acts as catalyst during conversion of complex sugars (sucrose) into glucose and fructose, which further react with another enzyme, zymase, to produce ethanol. The ethanol thus produced can be distilled, dehydrated and blended with gasoline. The reaction is exothermic and carbon dioxide is released in this process.

Enzymes play a central role in the production of ethanol from various substrates (biomass feed stocks) viz., wood, corn, straw, wheat, molasses, bagasse etc. In these feed stocks, the fermentable sugars are in the form of cellulose, which are surrounded by some polymers-lignin and hemicellulose. To solve this problem, the substrate needs to be pre-treated to expose the cellulose to a group of enzymes called cellulases. The use of high concentrations of cellulase accounts for more than 50% of ethanol production cost, which results in comparatively expensive process. The process is fast and very effective. Production of enzymes (from different sources) responsible for conversion process is important. Also, prospecting for new enzymes and enzyme producing organisms is also important. While production of ethanol, for making the process commercially viable, the effectiveness of the enzymes should be improved and simultaneously reducing the quantity needed.

Enzymes catalyze the chemical reactions by hydrolyzing biomass i.e., by converting complex sugars into ethanol. The process conditions, such as temperature, acidity, substrate concentration and many other process conditions influence the action of enzymes. The sugars present in the substrate can be converted into a number of chemical intermediates in biological reactors. Biological reactions are generally slow and selection of the most efficient reactor is important in these applications. Immobilized-cell reactors allow high cell densities and high throughput by attaching the microorganisms to a fixed support. In this line, an immobilized cell reactor was used for the production of ethanol in order to improve the performance of the fermentation process in the presence of *Saccharomyces cerevisiae. The yeast was immobilized by the enriched cells culture media harvested at exponential growth phase. The performances of the reactors of both batch fermentation as well as immobilized cells were compared along with the ethanol production. A rate model was also being proposed for performance of the two reactors (*Najafpour et al., 2004). *Similarly, the continuous production of ethanol from beet molasses by calcium alginate immobilized Saccharomyces cerevisiae in a packed-bed bioreactor was investigated. Compared to packed-bed bioreactor, lower ethanol concentration (3.94%) with productivity (8.67 g/L/h) was obtained in the continuous stirred bioreactor (Goksungur and Zorlu 2001).*

The rate of production of ethanol from glucose by *Saccharomyces cerevisiae* in a reactor needs to be studied. Ethanol production by reactors depend on many factors, viz., total reactor volume, its performance (can be influenced by nutrient and product transport rates), accumulation of carbon dioxide, process stability, etc. The future of ethanol would depend on reducing production costs and improving quality by fermentation, downstream processing and improvement in the enzymatic conversion processes.

ECONOMIC ASPECTS

Large scale fermenters are required for the production of ethanol. In a continuous operation for ethanol production, the economics of the production process i.e., economic factors to be taken into account during the fermentation process include substrate costs, fermenter conditions, operating costs, energy, investment, safety and market strategy. High production costs as well as technical difficulties are the possible disadvantages while using enzymes. On the other hand, the main advantage of using yeast directly during fermentation instead of enzymes includes its ability to grow at acidic pH.

SINGLE CELL PROTEIN

The physico-chemical characterization of preserved cashew apple juice used for the growth of *Saccharomyces cerevisiae* with an objective to produce single cell protein is presented in Table 4.

Table 4. Analysis of parameters for *Saccharomyces cerevisiae* grown in cashew apple juice

Fermentation time (h)	Parameter					
	pH	TSS (% Brix)	O.D (610 nm)	Protein (g/L)	Sugars (g/L)	Biomass (g/L)
0	4.96	8.2	0.05	0.05	0.53	0.00
3	3.46	11.9	0.17	0.14	0.49	0.44
6	3.08	13.2	0.15	0.20	0.44	0.71
16	2.87	14.7	0.98	0.28	0.21	1.115
24	3.01	12.8	0.8	0.22	0.18	1.09
30	3.45	12.5	0.78	0.18	0.18	0.83

From Table 3, it is observed that maximum values of optical density, total soluble solids and protein concentration were attained at 16 hours of fermentation. The decline in the growth of *Saccharomyces cerevisiae* NCIM 3211after 16 hours might be due to the exhaustion of the nutrients. During fermentation, the increase in the amount of biomass corresponds with the protein content. A maximum amount of biomass (1.115 g/L) was obtained at 16 hours of fermentation. The increase in biomass content with increase in fermentation time upto 16 hours might be due to the increased growth of the organism and the later decrease in biomass content could be attributed to the depletion of nutrients and accumulation of protein in the cells. At 16 hours, total soluble solids and sugar contents were also found to be high. From these results, it can be revealed that 16 hours is the optimum time of fermentation for maximum yield of single cell protein (0.28 g/L). Use of fresh cashew apple juice could have resulted in a higher yield, which could be used as a dietary supplement.

The decrease in the growth and low yield of single cell protein could be attributed to the limiting nutrients present in the juice as a result of prolonged preservation. Factors, such as batch of fruits taken for fermentation, percentage of mature fruits in a particular batch, etc. affect the composition of the substrate. The chemical composition of substrate should be taken into consideration, particularly the sugar content, which should be high enough, as required by the microbial strain for conversion into a desired product. Fermentation of cashew apple juice by *Saccharomyces cerevisiae* in a 7 litre batch fermenter at pH 5.0 and temperature 30 °C has resulted in a yield of 0.39 g cells/g reducing sugars (Layokun et al. 1986).

Researchers have produced single cell protein utilizing different substrates, such as pineapple, lemon pulp, papaya juice, among others. Maragatham and Paneerselvam (2011) produced 34 % of protein from papaya fruit extract using *Saccharomyces cerevisiae*. For the production of single cell protein, Rale (1984) used pineapple pulp for the growth of *Hansenula sydowiorum* and observed an average yield of 43.0 g/L cell biomass. de Gregorio *et al.,* (2002) utilised *Trichoderma viridae* for the production of single cell protein from lemon pulp and obtained 31.9 % nitrogen. From the above studies, it can be inferred that altering the chemical composition of the substrate by supplementing nutrients to the substrate would yield more single cell protein.

HYALUORONIC ACID

Feasibility of utilizing cashew apple juice as substrate for the production of hyaluronic acid from *Bacillus subtilis* MTCC 441was studied and the results obtained are presented in Table 5. The cell and hyaluronic acid concentrations have increased upto 28 hours of fermentation and later decreased.

Table 5. Cell concentration and hyaluronic acid content of fermented cashew apple juice

Fermentation time (h)	Cell concentration (O.D@600 nm)	Hyaluronic acid (g/L)
0	0.49	-
20	0.31	0.005
24	0.37	0.013
28	0.42	0.028
32	0.39	0.019

The maximum hyaluronic acid concentration obtained at the end of 28 hours by the fermentation of cashew apple juice with *Bacillus subtilis* was found to be 0.028 g/L. The optical density of the sample was found to be 0.42, indicating that *Bacillus* is a slow growing organism. Oliveira et al., (2005) reported a yield of 1,760 g/L from Streptococcal fermentation of cashew apple juice. Pires et al., (2010) used cashew apple juice for the production of hyaluronic acid by *Streptococcus zooepidemicus* and reported a maximum yield of 0.89 g/L.

Bacillus subtilis-derived material was secreted in high quality, which can be compared with commercially available sources of hyaluronic acid. Liu et al. (2008) produced hyaluronic acid by a two-stage culture strategy using *Streptococcus zooepidermicus* and observed an increase in the production of hyaluronic acid from 5.0 g/L to 6.6 g/L in batch culture. Widner et al. (2005) produced hyaluronic acid in *Bacillus subtilis* by expressing the hasA gene from *Streptococcus equisimilis*.

The large difference in the yield could be attributed to the use of different microbial strain and controlled environmental conditions. Another possible reason might be the variation in the physico-chemical quality of the fruits, region and climatic conditions of the fruits.

Despite the low yield, use of *Bacillus subtilis* for the production of hyaluronic acid was found to be advantageous in terms of simplified recovery process due to the secretion of the product into the medium. Additionally, the media required for the growth of *B.subtilis* was simple whereas the medium for streptococcal cultures are complex and expensive. Further, *B.subtilis*-derived hyaluronic acid would be exotoxin free, which is not possible in the case of certain streptococcal cultures, which could be problematic.

Higher yield might be obtained if the fermentation would have been carried out in bioreactor under controlled conditions. The growth was also affected by the amount of nutrients. Cashew apple juice supplied with micronutrients might increase the yield of hyaluronic acid.

CONCLUSION

From the literature available on cashew apples, it can be observed that large quantities of cashew apples are being wasted. Natural cashew apple juice is not utilized completely in India whereas it is partially utilized in the preparation of cashew apple juice derived products in Brazil. Owing to its high vitamin content, the utilization of cashew apples in various forms must be encouraged. An alternative to reduce this wastage is either to preserve the juice or to transform into value-added products. Microbial fermentation is the traditional way of production of such products but the use of enzymes might enhance the yield of these products.

The fresh cashew apple juice was utilised as fermentation medium for the production of ethanol, hyaluronic acid and preserved juice for the production of single cell protein. The studies carried out at shake flask level indicate that the fermentation of fresh juice with *Saccharomyces cerevisiae* NCIM 3190 yielded an ethanol concentration of 7.8 % at 26 hours whereas juice fermented with *Bacillus subtilis* yielded 0.028 g hyaluronic acid at 28 hours. The preserved cashew apple juice in the presence of *Saccharomyces cerevisiae* NCIM 3211 yielded 0.28 g/L single cell protein at 16 hours of fermentation. The product yields could have been improved further with nutrient supplementation under controlled conditions. Thus, the present work presents scope for rural farmers and labourers to earn remunerative returns for their produce; thereby generating better job opportunities, and improving their living standards from utilizing the agricultural waste that is otherwise discarded in the field.

ACKNOWLEDGMENTS

The authors are grateful to the Management, GITAM University and the Department of Biotechnology, GITAM Institute of Technology for extending their support during the implementation of the work.

REFERENCES

Akinwale, 1999. Fermentation and post fermentation chances in cashew wine. *J. Food Tech. Afr.* 4, 100-102.

Akinwale, T.O., Olubamiwa, O., Ajav, E.A. 2001. Cottage processing of cashew apple juice in Nigeria: Physico-chemical and sensory evaluation of product. *J. Food Tech. Afr.* 6, 56-58.

Araujo, S.M., Silva, C.F., Moreira, J.J.S., Narain, N., Souza, R.R. 2010. Biotechnological process for obtaining new fermented products from cashew apple fruit by *Saccharomyces cerevisiae* strains. *J. Ind. Microbiol. Biotechnol.* 38, 1161-1169.

Assuncao, R.B., Mercadante, A.Z. 2003b. Carotenoids and ascorbic acid from cashew apple (*Anacardium occidentale* L.): variety and geographic effects. *Food Chem.* 81, 495-502.

Attri, B.L. 2009. Effect of initial sugar concentration on the physico-chemical characteristics and sensory qualities of cashew apple wine. *Nat. Prod. Rad.* 8, 374-379.

Augustin, A. 1982. Studies on the clarification of cashew apple for the preparation of cashew syrup. *Indian J. Nutr. Diet.* 19, 169-172.

Azam-Ali, S.H., Judge, E.C. 2001. Small scale cashew nut processing. FAO, Rome. ITDG. Schumacher centre for Technology & Development. Bourton on Dunsmore, Rugby, Warwickshire, UK. Fao.org/Ag/ags/Agsi/Cashew/Cashew.htm (Accessed August 2009).

Azevedo, D.C.S., Rodrigues, A. 2000. Obtainment of high-fructose solutions from cashew (*Anacardium occidentale*) apple juice by simulated moving-bed chromatography. *Sep. Sci. Technol.* 35, 2561-2581.

Azevedo, D.C.S., Rodrigues, A. 2005. Separation of fructose and glucose from cashew apple juice by SMB chromatography. *Sep. Sci. Technol.* 40, 1761-1780.

Azoubel, P.M., Cipriani, D.C., El-Aouar, A.A., Antonio, G.C., Murr, F.E.X. 2005. Effect of concentration on the physical properties of cashew juice. *J. Food Engg.* 66, 413-417.

Bidaisee, G., Badrie, N. 2001. Osmotic dehydration of cashew apples (*Anacardium occidentale* L.): quality evaluation of candied cashew apples. *Int. J. Food Sci. Technol.* 36, 71-78.

Bopaiah, B.M. 1985. Microbial spoilage of cashew apples and its prevention. *Indian Cashew J.* 16, 15-17.

Campos, D.C.P., Santos, A.S., Wolkoff, D.B., Matta, V.M., Cabral, L.M.C., Couri, S. 2002. Cashew apple juice stabilization by microfiltration. *Desalination.* 148, 61-65.

Cavalcante, A.A.M., Rubensam, G., Erdtmann, B., Brendel, M., Henriques, J.A.P. 2005. Cashew (*Anacardium occidentale*) apple juice lowers mutagenicity of aflatoxin B1 in *S. typhimurium* TA102. *Genet. Mol. Biol.* 28, 328-333.

Chagas, C.M.A., Honorato, E.T., Pinto, E.G.A.S., Maia, E.G.A., Rodrigues, S. 2007. Dextransucrase production using cashew apple juice as substrate: effect of phosphate and yeast extract addition. *Bioproc. Biosys. Eng.* 30, 207-215.

Chempakam, B. 1983. Distribution of ascorbic acid and ascorbic acid oxidase activity in the developing cashew apple (*Anacardium occidentale* L.). J. Hortic. Sci. 58, 447-448.

Cormier, R. 2008. *Clarification of cashew apple juice and commercial applications.* Benin, West Africa.

Damasceno, L.F., Fernandes, F.A.N., Magalhaes, M.M.A., Brito, E.S. 2008. Evaluation and optimization of non enzymatic browning of "cajuina" during thermal treatment. *Braz. J. Chem. Eng.* 25, 313-320.

da Silva, K.D.P., Collares, F.P., Finzer, J.R.D. 2000. A simple and rapid method for estimating the content of solids in industrialized cashew juice. *Food Chem.* 70, 247-250.

de Carvalho, J.M., Maia, G.A., de Figueiredo, R.W., de Brito, E.S., Rodrigues, S. 2007. Storage stability of a stimulant coconut-cashew apple juice beverage. *J. Food Process. Pres.* 31, 178-189.

de Gregorio, A., Mandalari, G., Arena, F., Nucita, F., Tripedo, M.M., Curto, R.B.L. 2002. SCP and crude pectinase production by slurry-state fermentation of lemon pulps. *Biores. Technol.* 83, 89-94.

Fontes, C.P.M.L., Honorato, T.L., Rabelo, M.C., Rodrigues, S. 2009. Kinetic study of mannitol production using cashew apple juice as substrate. *Bioproc. Biosys. Engg.* 32, 493-499.

Giro, M.E.A., Martins, J.J.L., Rocha, M.V.P., Melo, V.M.M., Goncalves, L.R.B. 2009. Clarified cashew apple juice as alternative raw material for biosurfactant production by *Bacillus subtilis* in a batch bioreactor. *Biotechnol. J.* 4, 738-747.

Goksungur, Y., Zorlu N. *2001.* Production of ethanol from beet molasses by ca-alginate immobilized yeast cells in a packed-bed bioreactor. *Turk. J. Biol.* 25, 265-275.

Gunjate, R.T., Patwardhan, M.V. 1995. Cashew. In: Salunkhe D.K, Kadam S.S (Eds.), *Handbook of Fruit Science and Technology: production, composition, storage and processing*, CRC, USA. 509-526.

Honorato, T.L., Rodrigues, S. 2008. Dextransucrase stability in cashew apple juice. *Food Bioproc. Technol.* 3, 105-110.

Joseph, A.D.O. 2010. Comparative studies of wine produced by spontaneous and controlled fermentation of preserved cashew (*Anacardium occidentale*) juice. *Res. J. Biol. Sci.* 5, 460-464.

Karuppaiya, M., Sasikumar, E., Viruthagiri, T., Vijayagopal, V. 2009. Optimization of process parameters using response surface methodology (RSM) for ethanol production from waste cashew apple juice using *Zymomonas mobilis*. *Chem. Eng. Comm.* 196, 1-11.

Karuppaiya, M., Sasikumar, E, Viruthagiri, T., Vijayagopal, V. 2010. Optimization of process variables using response surface methodology (RSM) for ethanol production from cashew apple juice by *Saccharomyces cerevisiae*. *As. J. Food Ag-Ind.* 3, 462-473.

Layokun, S.K., Obawole, A.B., Fatile, I.A., Solomon, B.O. 1986. Investigation of cashew apple juice as a substrate for single cell protein production. *J. Food Sci.* 51, 237-238.

Liu, L., Du, G., Chen, J., Wang, M., Sun, J. 2008. Enhanced hyaluronic acid production by a two-stage culture strategy based on the modeling of batch and fed-batch cultivation of *Streptococcus zooepidemicus*. *Biores. Technol.* 99, 8532-8536.

Lowor, S.T., Agyente-Badu, C.K. 2009. Mineral and proximate composition of cashew apple (*Anacardium occidentale* L.) juice from Northern Savannah, forest and Coastal Savannah regions in Ghana. *Am. J. Food Technol.* 4, 154-161.

Maciel, M.I., Hansen, T.J., Aldinger, S.B., Laboes, J.N. 1986. Flavour chemistry of cashew apple juice. *J. Agric. Food Chem.* 34, 923-927.

Maragatham, C., Panneerselvam, A. 2011. Production of single cell protein from yeast using papaya extract medium. *Adv. Appl. Sci. Res.* 2, 14-18.

Marc, A., Achille, T.F., Mory, G., Koffi, P.V.N., Georges, A.N. 2011. Minerals composition of the cashew apple juice (*Anacardium occidentale* L.) of Yamoussoukro, Cote D'ivoire. *Pak. J. Nutr.* 10, 1109-1114.

Mandal, R.C., Dhandar, D.G., Dhume, C.V. 1985. Cashew apple liquor industry in Goa. *Acta Hort. (ISHS).* 108, 172-175.

Michodjehoun-Mestres, L., Souquet, J-M., Fulcrand, H., Bouchut, C., Reynesa, M., Brillouet, J-M. 2009. Monomeric phenols of cashew apple (*Anacardium occidentale* L.). *Food Chem.* 112, 851–857.

Mohanty, S., Ray, P., Swain, M.R., Ray, R.C. 2006. Fermentation of cashew (*Anacardium occidentale* L.) "apple" into wine. *J. Food. Proc. Pres.* 30, 314-322.

Morton, J.F. 1961. The cashew's brighter future. *Economic Botany.* 15, 57-78.

Morton, J.F., Dowling, C.F. 1987. Cashew Apple. In: Miami F.L (Ed.), *Fruits of Warm Climates*, Creative Resources Systems, Florida. 239-240.

Muir-Beckford, M., Badrie, N. 2000. Consumer acceptance of tropical wines from *Aloe vera* (*Aloe barbadensis*) and cashew apples (*Anacardium occidentale* L.) in the British Virgin Islands. *Foodservice Res. Int.* 12, 185-196.

Muroi, H., Aya, K., Isao, K. 1993. Antimicrobial activity of cashew apple flavour compounds. *J. Agric. Food Chem.* 41, 1106-1109.

Najafpour, G., Yousnesi, H., Ismail, K.S.K. 2004. Ethanol fermentation in an immobilized cell reactor using *Saccharomyces cerevisiae*. *Biores. Technol.* 92, 251-260.

Neelakandan, T., Usharani, G. 2009. Optimization and production of bioethanol from cashew apple juice using immobilized yeasts cells by *Saccharomyces cerevisiae*. *American-Eurasian J. Sci. Res.* 4, 85-88.

Oduwole, O.O., Akinwale, T.O., Olubamiwa, O. 2001. Economic evaluation of a locally fabricated extraction machine for a cottage cashew juice factory. J. *Food Technol. Afr.* 6, 18-20.

Oliveira, A.H.S., Ogrodowski, C.S., Goncalves, L.R.B., Santana, M.H.A. 2005. Natural cashew apple juice as fermentation medium for hyaluronic acid production by *Streptococcus zooepidemicus*. *Proc. 27th Symp on Biotechnology for fuels and chemicals*, Colorado, 1-4 May, 152.

Osho, A. 1995. Evaluation of cashew apple juice for single cell protein and wine production. *Food / Nahrung.* 39, 521–529.

Pinheiro, A.D.T., Rocha, M.V.P., Macedo, G.R., Goncalves, L.R.B. 2008. Evaluation of cashew apple juice for the production of fuel ethanol. *Appl. Biochem. Biotechnol.* 148, 227-234.

Pires, A.M.B., Macedo, A.C., Eguchi, S.Y., Santhana, M.H.A. 2010. Microbial production of hyaluronic acid from agricultural resource derivatives. *Biores. Technol.* 101, 6506-6509.

Price, R.L., Holanda, L.F.F., Fe, J.A.M., Maia, G.A., Martins, C.B. 1975. Constituents of Brazilian cashew apple juice. *Cien. Agron.* 5, 61-65.

Rabelo, M.C., Fontes, C.P.M.L, Rodrigues, S. 2009. Enzyme synthesis of oligosaccharides using cashew apple juice as substrate. *Biores. Technol.* 100, 5574-5580.

Rale, V.B. 1984. SCP from pine apple (*Ananas sativa* Schutt) Cannery effluents. *Appl. Microbiol. Biotechnol.* 19, 106-109.

Rale, V.B. 1985. Production of feed yeast and yeast- enriched livestock feed from cashew apple. *Mircen J.* 1, 205-212.

Rocha, M.V.P., Oliveira, A.H.S., Souza, M.C.M., Goncalaves, L.R.B. 2006. Natural cashew apple juice as fermentation medium for biosurfactant production by *Acinetobacter calcoaceticus. World J. Microbiol. Biotechnol.* 22, 1295-1299.

Silva, M.E., Torres Neto, A.B., Silva, W.B., Silva, F.L.H., Swarnakar, R. 2007. Cashew wine vinegar production: alcoholic and acetic fermentation. *Braz. J. Chem. Eng.* 24, 163-169.

Silveira, M.S., Fontes, C.P.M.L., Guilherme, A.A., Fernandes, F.A.N., Rodrigues, S. 2010. Cashew apple juice as substrate for lactic acid production. *Food Bioprocess Technol.* 5, 947-953.

Talasila, U., Vechalapu, R.R, Shaik, K.B. 2011. Preservation and shelf life extension of cashew apple juice. *Internet J. Food Safety.* 13, 275-280.

Thiripurasundari, G., Usharani, G. 2011. Comparative production of vinegar using cashew apple juice by different immobilization techniques. *Curr. Bot.* 2, 31-33.

Trevisan, M.T.S., Pfundstein, B., Haubner, R., Wurtele, G., Spiegelhalder, B., Bartsch, H., Owen, R.W. 2006. Characterization of alkyl phenols in cashew (*Anacardium occidentale*) products and assay of their antioxidant capacity. *Food Chem. Toxic.* 44, 188-197.

Vergara, C.M.A.C., Honorato, T.L., Maia, G.A., Rodrigues, S. 2010. Prebiotic effect of fermented cashew apple (*Anacardium occidentale* L) juice. *Food Sci. Technol.* 43, 141-145.

Widner, B., Behr, R., Dollen, S.V., Tang, M., Heu, T., Sloma, A., Sternberg, D., DeAngelis, P.L., Weigel, P.H., Brown, S. 2005. Hyaluronic acid production in *Bacillus subtilis. Appl. Environ. Microbiol.* 71, 3747-3752.

In: Enzymes in Value-Addition of Wastes
Editors: Satinder Kaur Brar and Mausam Verma
ISBN: 978-1-62948-968-1

Chapter 10

KEY ENZYMES IN VALUE-ADDITION OF CRUDE GLYCEROL TO BIOHYDROGEN

***Saurabh Jyoti Sarma*[1], *Gurpreet Singh Dhillon*[1], *Satinder Kaur Brar*[1*], *Yann Le Bihan*[2] *and Gerardo Buelna*[2]**

[1]Institut National de la Recherche Scientifique (INRS)
Centre Eau, Terre & Environnement, Québec(QC), Canada
[2]Centre de recherche industrielle du Québec (CRIQ), Québec(QC), Canada

ABSTRACT

Crude glycerol is a byproduct of biodiesel production by transesterification of vegetable oils and animal fats. Nearly 10 Kg of crude glycerol is produced for every 100 Kg of biodiesel produced by transesterification process. Corresponding to huge increase in global biodiesel production, crude glycerol production is also increasing in recent years. As crude glycerol is very cheap and a good source of nutrient for microbial growth, it can be considered as a potential feedstock for industrial fermentation. Hydrogen is an emerging biofuel with very high energy content and greenhouse gas reduction potential. Numbers of recent studies have suggested that crude glycerol is a suitable substrate for microbial hydrogen production. Dark fermentation and photo-fermentation are the two processes which have been mostly investigated for fermentative hydrogen production. For both the processes, biochemical pathways have been worked out and enzymes associated with such pathways have been characterized. In the present chapter, the key enzymes involved in the crude glycerol bioconversion processes for hydrogen production has been identified and their structural and biochemical properties have been discussed. Further, present research activities on manipulation of the enzymatic pathways for increased hydrogen yield have also been summarized. Similarly, further research possibilities on the enzyme system involved in crude glycerol bioconversion process have been identified. Thus, the purpose of this chapter is to have a better understanding of the biochemical reactions and the corresponding enzymes responsible for biohydrogen production, and recent advances in this field.

[*] Corresponding author, Phone: 1 418 654 3116; Fax: 1 418 654 2600; E-mail: satinder.brar@ete.inrs.ca.

Keywords: Biodiesel, crude glycerol, enzyme, hydrogenase, hydrogen production

1. INTRODUCTION

Hydrogen is a clean energy source and considered to be the most promising energy carrier for future [1, 2]. Presently, industrial production of hydrogen is dominated by steam reforming of fossil fuels and electrolysis of water [1, 2]. Alternatively, hydrogen production by fermentation of different organic substrates is an environmentally sustainable method for hydrogen production [1-3]. Different feedstock ranging from agricultural waste biomass to industrial organic wastes is known to be used for biological hydrogen production [4, 5]. Crude glycerol (CG) generated during biodiesel production process is a relatively new substrate evaluated for microbial hydrogen production [6]. Actually, during biodiesel production by transesterification of vegetable oil and animal fat, for every 100 Kg of biodiesel, 10 kg of CG is generated [7, 8]. Further, global biodiesel production is rapidly increasing [9] and hence, global CG production will also increase in coming years. Moreover, due to the presence of different impurities, CG has negligible commercial importance and as a consequence, considering its unavoidable increased production, management of this waste may be a concern for biodiesel manufacturers. Alternatively, if it could be used as a feedstock for commercial production of bio-hydrogen it may have dual advantages; firstly, biodiesel manufacturers will be economically benefited and secondly, produced biofuel (H_2) will help in greenhouse gas reduction by replacing fossil fuels. Dark fermentation [10, 11], photo-fermentation [12] and bio-electrochemical/microbial electrolysis cell [13, 14] are the three different methods so far tested for hydrogen production by CG bioconversion. Similarly, microorganisms such as, facultative anaerobic bacteria (*Enterobacter aerogenes, Klebsiella pneumoniae*), photosynthetic bacteria (*Rhodopseudomonas palustris*), hyperthermophilic eubacterium (*Thermotoga neapolitana*) or mixed microbial cultures are known to be used for this purpose [10-12, 15, 16]. From the above discussion, it can be concluded that till now, a range of physiologically different microorganisms have been evaluated for CG bioconversion and hydrogen production. However, for a particular microorganism, thermodynamic limitations of substrate utilization and product formation are precisely determined by its enzyme systems and metabolic pathways. Therefore, detailed information on the biochemical pathways and key enzymes involved in hydrogen production and CG bioconversion will be helpful for improved utilization of the process. Thus, the purpose of this chapter is to discuss the enzymatic pathways, biochemical properties of key enzymes and recent advances in metabolic/enzyme engineering in the particular fields.

2. IMPORTANT METABOLIC PATHWAYS

2.1. Glycerol Bioconversion Pathways

Glycerol bioconversion pathways can be differentiated in to oxidative and reductive pathways. Different intermediate metabolites and enzymes involved in both the pathways are shown in Figure 1. In oxidative bioconversion of glycerol, the first reaction is initiated by glycerol dehydrogenase where glycerol is converted to dihydroxyacetone. This intermediate is subsequently phosphorylated by dihydroxyacetone kinase. The phosphorylated intermediate finally enters glycolysis [17, 18], where in anaerobic dark fermentation sets in, and depending upon the microorganism used, it is further metabolized to H_2, CO_2 and different fermentation end-products. Therefore, from H_2 production point of view, oxidative metabolism of glycerol is preferred. Unlike, the oxidative branch, reductive metabolism of glycerol is simpler. In this pathway, glycerol is first converted to an intermediate 3-hydroxypropionaldehyde by the catalytic activity of glycerol dehydratase, a vit-B_{12}-dependent enzyme and diol dehydratases. The final step of this pathway is the conversion of 3-hydroxypropionaldehyde to 1,3-propanediol by the catalytic activity of $NADH^+H^+$- dependent 1,3-propanediol dehydrogenase [18-20]. Interestingly, both oxidative and reductive branch of the metabolic pathway of glycerol may be observed in *Enterobacter, Klebsiella, Clostridium* and *Citrobacter* [18]. In such cases, presence of both the pathways and simultaneous production of H_2 and 1, 3-propanediol may be responsible for poor H_2 yield per mole glycerol consumed.

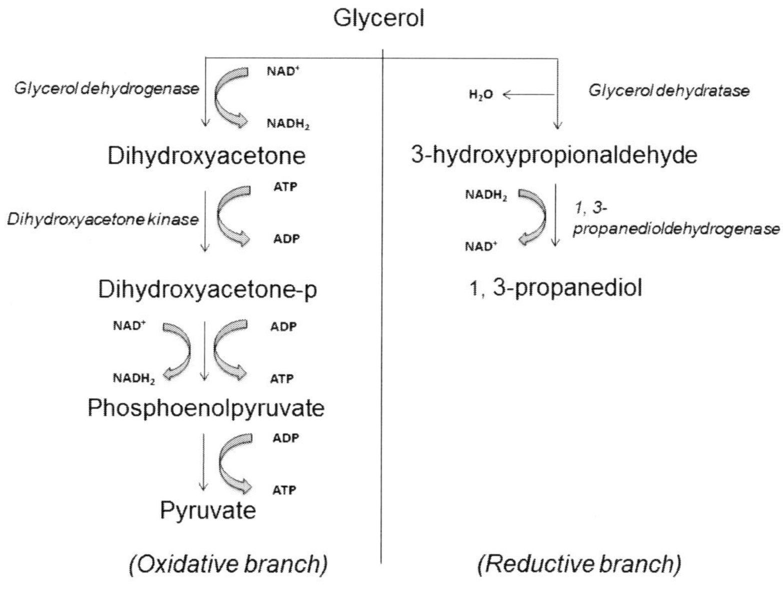

Figure 1. Schematic representation of the glycerol bioconversion pathways (Adapted from [6, 17, 18]).

2.2. Hydrogen Production Pathways

There are three different well known metabolic pathways for fermentative hydrogen production. These comprise:

1. Pyruvate: ferredoxin oxido reductase (PFOR) pathway[21, 22]
2. Pyruvate: formate lyase (PFL) pathway[22, 23]
3. Pentose phosphate (PP) pathways [24]

More than one such pathway may be present in some organisms; however, only one pathway is generally dominant in hydrogen production [25]. Each of these pathways is separately discussed in the succeeding sections.

2.2.1. Pyruvate: Ferredoxin Oxido Reductase (PFOR) Pathway

PFOR pathway is the characteristic of strict anaerobic hydrogen producers, such as *Clostridium* [25]. Organisms of this group include *Clostridium butyricum* which ferments carbohydrates to produce H_2 accompanied by CO_2, acetate and butyrate [21, 26-28]. As shown in Figure 2, under the catalytic activity of pyruvate: ferredoxin oxidoreductase (PFOR), pyruvate is broken down to acetyle-CoA and CO_2 [21, 22]. A portion of the electrons generated during production of acetyle-CoA is transferred to ferredoxin to reduce it [22] and the other to NAD^+ to produce $NADH_2$ [21]. From reduced ferredoxin, the electrons are transferred to H^+ to produced H_2 and the reaction is catalyzed by hydrogenase [21]. Number of [FeFe] hydrogenases can reoxidize the reduced ferredoxin for hydrogen production and ferredoxin-dependent hydrogenase (Fd-[FeFe]) is one of them [22]. Further, NADH can also be utilized in hydrogen production and there are three possibilities:

a) It may produce hydrogen by reducing ferredoxin in the presence of NADH-ferredoxin oxidoreductase (NFOR),
b) H_2 may be directly produced from NADH by the catalytic activity of NADH-dependent [FeFe] hydrogenase (NADH-[FeFe]). However, hydrogen production from NADH is thermodynamically favorable only at low hydrogen partial pressure [29, 30], hence, the enzyme NADH-dependent [FeFe] hydrogenase may not be involved in hydrogen production in batch culture, where hydrogen accumulates in the headspace to substantial concentrations [30, 31].
c) H_2 may be produced by NADH- ferredoxin-dependent hydrogenase (Fd-NADH-[FeFe]) where NADH is utilized as a co-substrate with reduced ferredoxin [22].

In all the cases the balance NADH is utilized to produce different reduced fermentation end- products [22]. In the case of strict anaerobes, such as *Clostridium butyricum*, besides hydrogen, the fermentation end-products of anaerobic glucose metabolism are acetate and butyrate accompanied by ATP, biomass and reducing equivalents [21]. As already described, NADH can be accessed and more than one hydrogenase may be involved in hydrogen production by PFOR pathway. Therefore, depending upon the hydrogenase used, hydrogen yield may vary from 2–4 mol of hydrogen per mole of glucose [22]. This is the advantage of PFOR pathway over pyruvate: formate lyase (PFL) pathway as the latter is not able to utilize NADH for hydrogen production and maximum hydrogen yield is only 2 mol hydrogen per mole of glucose.

2.2.2. Pyruvate: Formate Lyase (PFL) Pathway

Facultative anaerobes, such as, *E. coli* and *Enterobacter* species use PFL pathway for hydrogen production [22, 23, 32]. A simplified PFL pathway is shown in Figure 3. As it can be seen from the Figure 3, in PFL pathway, unlike PFOR pathway, pyruvate is broken down to acetyl-CoA and formate [22].

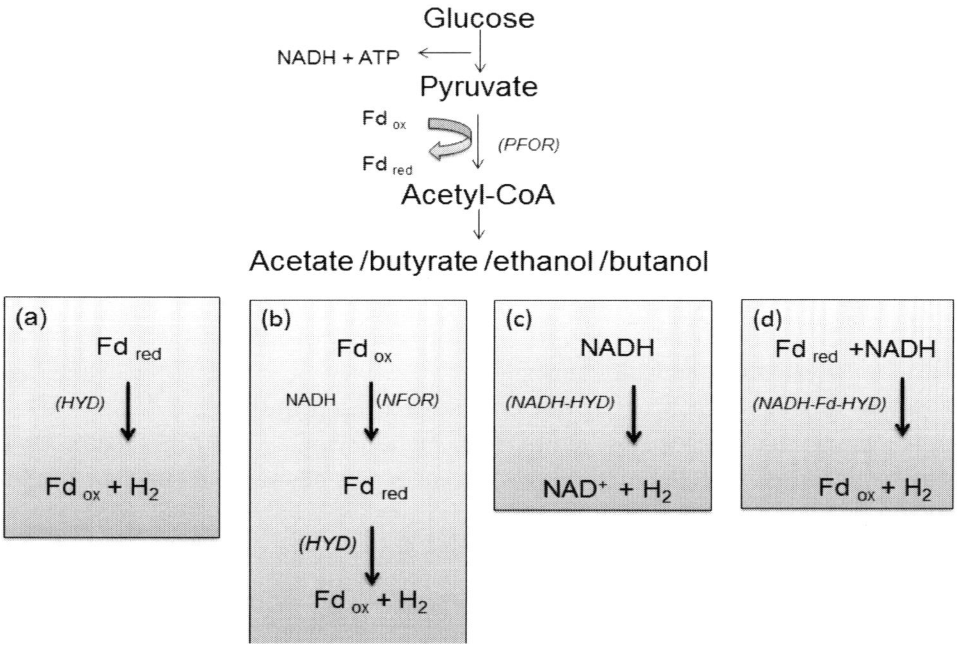

Figure 2. Simplified representation of PFOR pathway for hydrogen production. (a-d) shows 4 different mechanisms involved in hydrogen evolution where, PFOR= Pyruvate: ferredoxin oxido reductase, Fd $_{ox/red}$= Oxidized or reduced ferredoxin, HYD= (hydrogenase), NADH-HYD= (NADH-dependent [FeFe] hydrogenase), NADH-Fd-HYD= (NADH- ferredoxin-dependent hydrogenase) and NFOR= (NADH-ferredoxin oxidoreductase) (adapted from [21, 22, 33]).

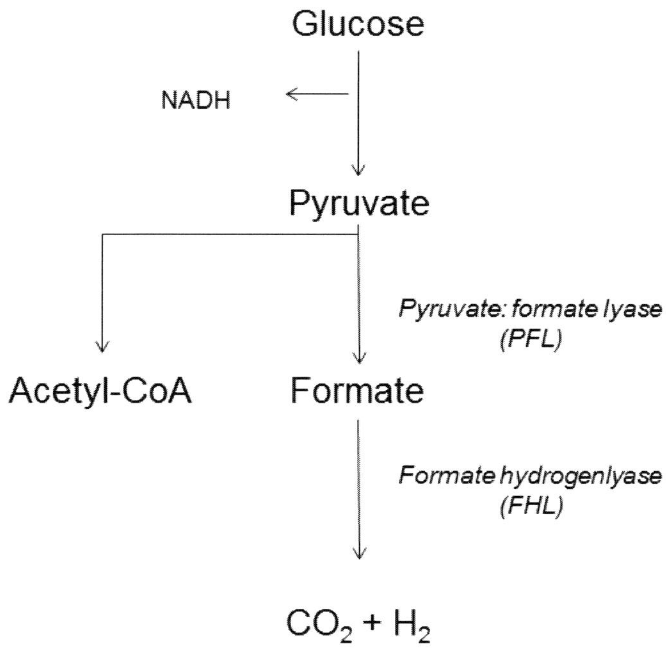

Figure 3. Simplified representation of PFL pathway for hydrogen production (adapted from [22, 34]).

Further, there are two different mechanisms of utilizing formate for hydrogen production:

a) In the [NiFe] hydrogenase dependent mechanism, formate is broken down to CO2 and H2 by the catalytic activity of multicomponent, membrane-associated enzyme complex FHL (formate hydrogenlyase) [32].
b) In some other organisms, H_2 production is realized by formate dependent [FeFe] hydrogenase [22].

2.2.3. Pentose Phosphate (PP) Pathway

NAD(P)H-dependent pentose phosphate (PP) pathway is the most efficient metabolic pathway for hydrogen production using dark fermentation [24]. Theoretically, higher hydrogen yields are possible when hexose conversion takes place via pentose phosphate (PP) pathway instead of the Embden–Meyerhof (EM) and Entner–Doudoroff (ED) pathways [35]. This pathway has the potential to produce stoichiometric amount of H_2 per mole of glucose [36, 37].

Hexose monophosphate shunt or pentose phosphate pathway (PPP) is an alternative pathway for glucose oxidation occurs in the cytosol of all cells. PPP can function in the absence of oxygen and no ATP is generated during the process[38]. It generates the ribose 5-phosphate required for the synthesis of different nucleotides and the reducing agent nicotinamide adenine dinucleotide phosphate (NADPH) required for different metabolic reactions [38, 39]. In most animal tissues glucose oxidation mainly occurs via glycolysis [80–90%) however, the rest of the oxidation (10–20%) occurs via the PPP [38]. The PPP has two branches, one is oxidative branch and the other is non-oxidative branch [38-41]. The reactions of the oxidative branch are non-reversible whereas, the non-oxidative branch is reversible [38, 40]. A simplified PPP has been shown in Figure 4. As shown in Figure 4, the PPP pathway is interlinked with glycolysis and the flow of glycolysis intermediate such as, glucose-6-phosphate through the PPP depends on the metabolic demand of NADPH and ribose-5-phosphate [38]. There are three irreversible reactions in the oxidative branch of PPP and in the first reaction, glucose-6-phosphate, the glycolysis intermediate, undergoes dehydrogenation to form 6-phosphoglucono-d-lactone and NADPH [38]. This reaction is catalyzed by glucose 6-phosphate dehydrogenase (G6PDH) [38, 40] and under physiological conditions it is the rate limiting step of the pathway [41]. Further, by the activity of specific lactonase, 6-phosphoglucono-d-lactone undergoes rapid hydrolysis to produce 6-phosphogluconate. Similarly, 6-phosphogluconate undergoes oxidative decarboxylation to generate CO_2, NADPH and ribose-5-phosphate and the reaction is catalyzed by 6-phosphogluconate dehydrogenase (6PGD) [38, 39].

Furthermore, non-oxidative branch of pentose phosphate pathway can utilize fructose-6-phosphosphate as well as glyceraldehyde-3-phosphate, the glycolytic intermediates to produce the ribose-5-phosphate. These reactions are reversible and no NADPH is generated during the process. As the reactions are reversible, glyceraldhyde-3-phosphate can be reproduced from ribose-5-phosphate which can be further converted back to another glycolytic intermediate glucose-6-phosphate (Figure 4) [40]. Moreover, glucose-6-phosphate can re-enter the oxidative branch of PPP and can take part in NADPH generation. By proper manipulation of the two branches of PPP, six molecules of NAPDH can be generated in the expense of one molecule of glucose [40, 42, 43]. Actually, under certain circumstances, pentose phosphate pathway complete oxidization of glucose to CO_2 and water is possible

[44]. In a significant development, different enzymes of oxidative PPP have been reported to be successfully applied to produce 11.6 mol of H_2 per mol glucose-6-phosphate [45].

3. HYDROGENASES

Hydrogenases are metalloproteins and they were first discovered by Stephenson and Stickland in 1931 [46-49]. It may be present in different organisms ranging from bacteria (methanogenic, acetogenic as well as nitrate and sulfate reducing), anaerobic archaea, fungi, protozoa, rhizobia and some algae [48]. Most of the hydrogenases have Fe-S clusters and metal atom containing active center, where the atoms may be a Ni and an Fe atom or two Fe atoms [49]. These enzymes are responsible for reversible oxidation of hydrogen gas and directly involved in H_2 production (reduction of H^+) or H_2 consumption (oxidation of H_2) [47, 49]. Further, the direction of the reaction depends on the redox potential of the components interacting with the enzyme [47, 49]. Moreover, there is a certain hydrogenase found in autotrophic proteobacteria such as, *Ralstonia eutropha* and *Rhodobacter capsulatus* which acts as a sensor for intracellular hydrogen status and regulates the gene expression for hydrogenase production [47-51].

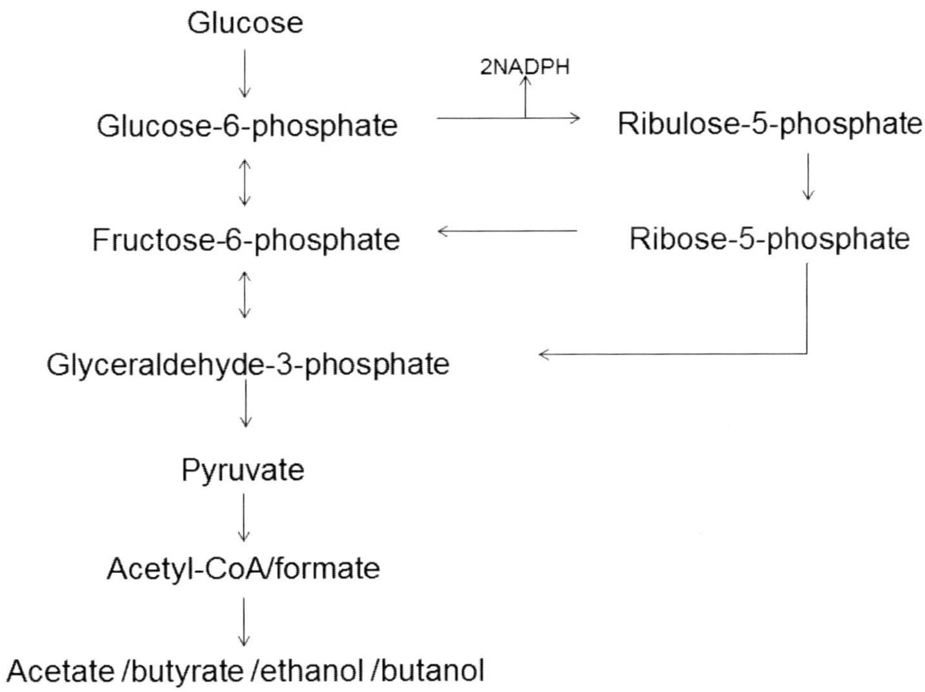

Figure 4. Simplified representation of pentose phosphate pathway for hydrogen production (adapted from [24, 52].

There is a great diversity among the hydrogenase with respect to their size, structure and metal containing active center. However, they can be broadly categorized in to three groups, such as [NiFe]- hydrogenase, [FeFe]- hydrogenase and [Fe]-hydrogenase and it is supported by the sequence information of different hydrogenases [53] [47]. Further, there is a transition metal free hydrogenases identified in methanogenic archaea [48, 54]. This group of htdrogenase is abbreviated as Hmd hydrogenase and its catalytic properties are different from the common hydrogenases [49, 55]. Further, unlike [NiFe]- and [FeFe]- hydrogenase, it does not contain any Fe-S cluster, however, it has an iron-containing cofactor [49].

3.1. [NiFe]-hydrogenases

[NiFe]-hydrogenases family is the most abundant among different types of hydrogenase [47]. However, [NiFe]-hydrogenases are relatively less active in hydrogen-uptake or production than Fe-hydrogenases Albracht 1994. They are generally involved in hydrogen uptake [48]. Different microorganisms can use hydrogen as an energy source and [NiFe]-hydrogenases mediate the cleavage of H_2 to allow the organism to obtain the required energy [50, 56]. Hence, this enzyme may be found in a diverse group of microorganisms. They may be the aerobic bacteria that oxidize H_2 to obtain energy, which may be either facultative or obligate chemolithotrophs as well as may be either gram-negative or gram-positive (Friedrich and Schwartz 1993. Further, they may be cyanobactria [57], methanogenic Archaea [58], phototrophic purple sulfur bacterium such as, *Thiocapsa roseopersicina* [59] as well as bacteria, such as *E. coli*. Hydrogenase 1 and hydrogenase 2 are the two hydrogen-oxidizing [NiFe]-hydrogenases, which *E. coli* utilizes during its anaerobic metabolism [60].

In general, [NiFe]-hydrogenase has a large subunit bearing the active site and a small subunit having Fe-S clusters which take part in electron transfer [47, 50, 61-63]. The active site of the enzyme contains the Ni and Fe atoms where the Fe atom is coordinated with one CO and two CN ligands [47, 64]. [NiFe]-hydrogenase has two oxidized states, Ni-A and Ni-B, where the former exhibits a lag phase in reductive activation [65]. Further, based on the sequence analysis of different subunits, [NiFe]-hydrogenases can be broadly classified in to four different groups and it is elaborately discussed by Kim and Kim (2011). In brief, the groups are (i) Group I: membrane-bound [NiFe] uptake hydrogenases, (ii) Group II: cytoplasmic H_2 sensing hydrogenase and cyanobacterial uptake [NiFe]-hydrogenases, (iii) Group III: reversible multimeric cytoplasmic [NiFe]-hydrogenases, where the additional subunits help the enzyme to bind with different co factors such as, NAD or NADP, (iv) Group IV: membrane associated hydrogenases with six or more subunits that are involved in H_2 production. These enzymes can use reduced ferredoxin for H_2 production and are mostly found in Archaea and also in *E. coli* [47].

Carbon monoxide can compete with hydrogen and inhibit the activity of [NiFe]-hydrogenase involved in hydrogen metabolism Albracht 1994. Similarly, presence of O_2 also has negative effect on the activity of this enzyme Albracht 1994. Actually, presence of oxygen can bring reversible inhibition of the enzyme [66]. However, it is known that [NiFe]-hydrogenases are less sensitive than [FeFe]-hydrogenases towards CO and O_2 inhibition [47, 67]. Based on X-ray crystallographic investigation on the [NiFe]-hydrogenase from *Desulfovibrio vulgaris* Miyazaki F, researchers have concluded that carbon monoxide binds to the nickel ion present in the active site of the enzyme to inhibit its catalytic activity [66, 68]. At temperatures below 100 K, upon illumination, reversible cleavage of the Ni-CO bonds is possible which can rebind in the absence of light [66]. However, in a more recent investigation, it is concluded that CO binds with Fe instead of Ni center of a [NiFe]-hydrogenase to inhibit its activity [69].

Similarly, acetylene is another gaseous inhibitor which can inhibit the activity of [NiFe]-hydrogenase and the phenomenon is observed in different organisms, such as *Azotobacter chroococcum*, *Desulfovibrio gigas* and *D. baculatus* [62, 70]. It is known that acetylene is capable of binding the large subunit of the *Azotobacter vinelandii* hydrogenase to inhibit its activity [62, 71]. Interestingly, acetylene binding can protect the enzyme from harmful effect of O_2, at the same time H_2 can prevent the binding of acetylene with the enzyme [62].

3.2. [FeFe]-hydrogenases

[FeFe]-hydrogenases are the key enzymes in hydrogen metabolism and found in a range of microorganisms, however, they are not as common as [NiFe]-hydrogenases, where the latter is present in almost all archaea and bacteria [72]. [FeFe]-hydrogenases are mostly found in different anaerobic prokaryotic organisms, such as clostridia [73, 74], and sulfate reducing bacteria [75]. Further, sulfate reducing bacteria are among the few organisms where both [NiFe]- and [FeFe]-hydrogenases are known to occur [76]. Moreover, [FeFe]-hydrogenases are also found in eukaryotic organisms, such as green algae [77]. Actually, they are the only enzymes of this class found in eukaryotes [47, 49, 78]. Among the transition metals containing hydrogenases, Fe-containing hydrogenases are an important class. These Fe-only hydrogenases are the enzymes that can catalyze the formation and cleavage of molecular hydrogen; however, they are usually involved in hydrogen production [48]. Molecular structure of [FeFe]-hydrogenases is different from that of the [NiFe]-hydrogenases, where, the former is monomeric and the latter has two different structural subunits. The active site of [FeFe]-hydrogenase bears the two Fe atoms and it is present in the conserved domain of the molecule, whereas, there may be an additional domain containing the Fe-S clusters. Further, the two Fe atoms are attached with CN and CO ligands and coordinated with two sulfur atoms [47, 79]. FeFe]-hydrogenases are mainly known for their involvement in H_2 production, although, their role in hydrogen uptake has also been reported in some organisms such as, *Desulfovibrio vulgaris* [47]. Hydrogen production rate of these enzymes are known to be 10 to 1000 times faster than [NiFe]-hydrogenases [80, 81]. Compared to [NiFe]-hydrogenases, they have 100 times less affinity for hydrogen, however, they have more sensitivity towards O_2 and CO inhibition [81].

There is one more class of iron-hydrogenase, which was formerly known as ''metal free hydrogenase'' and also familiar as Fe-hydrogenase [82]. These are the cytoplasmic enzymes and found in some members of methanogenic archaea [82]. These enzymes are structurally different from the [NiFe]- and [FeFe]-hydrogenases and unable to catalyze the production of hydrogen [47]. By consuming H_2 and CO_2, they can produce CH_4 [83, 84], and hence, they are not expected in a hydrogen production system.

4. NITROGENASES

The nitrogenase is the sole enzyme-complex involved in atmospheric nitrogen fixation. The activity is carried out by diazotrophic microorganisms at relatively mild temperatures [85]. Actually, nitrogen fixation is the characteristics of diazotrophic microorganisms, all of which are prokaryotes, either bacteria or archaea and widely distributed in the nature [85-87]. Such microorganisms may be a strict anaerobe (*Clostridia*), facultative anaerobe (*Klebsiella pneumoniae*), aerobe (*Azotobacter vinelandii*), oxygenic photosynthetic bacteria (*Anabaena cylindrica*) or anoxygenic photosynthetic bacteria (*Rhodobacter sphaeroides, Rhodopseudomonas palustris, Rhodobacter capsulatus*). During nitrogen fixation process, atmospheric nitrogen is reduced to ammonia by the activity of nitrogenase and the required energy is provided by ATP. For each mole of nitrogen fixed, 16 ATPs are consumed and hence, energetically it is one of the most expensive metabolic processes [87]. Interestingly, during this process protons are also obligatorily reduced to hydrogen. In fact, in a pure nitrogen environment, around 75% of electrons are used in reducing nitrogen to produce ammonia and the rest of electrons are available for hydrogen production [88]. Further, in the absence of nitrogen, more and more electrons are available for proton reduction and the nitrogenase acts as an ATP-driven hydrogenase [47, 89], generating exclusive amount of H_2 with negligible nitrogen fixation activity. Alternatively, with response to ammonium availability, synthesis and activity of nitrogenase are precisely regulated by multilevel feedback control systems [90] and hence, H_2 production is also regulated. Three different types of nitrogenases in the found in diazotrophic microorganisms and they are known as: (i) molybdenum nitrogenase, (ii) vanadium nitrogenases and; (iii) iron-only nitrogenases [47, 85]. Actually, this classification is based on the metal atoms present in the active site of the enzyme. *Rhodopseudomonas palustris* CGA009 is known to be the only photosynthetic bacterium which possesses the three kinds of nitrogenases. Among these enzymes, molybdenum nitrogenase are found in all diazotrophs and involved in most nitrogen fixation activity [85] and other nitrogenases are also known as alternative nitrogenases. Further, the expression of these alternative nitrogenases has a correlation with the availability of nitrogen [47]. Molybdenum nitrogenase is comprised of two proteins, one is dinitrogenase which contains two FeMo cofactors and the other is dinitrogenase reductase [47, 88]. Vanadium or iron nitrogenases also have similar structure to that of molybdenum nitrogenase, however, the cofactors are FeVa and FeFe, respectively [47]. From hydrogen production point of view, compared to molybdenum nitrogenase, alternative nitrogenases can use higher amount of electrons for proton reduction and high hydrogen yield is possible [47]. Nitrogenase is an oxygen sensitive enzyme and it can be irreversibly inactivated by O_2 [88, 91]. All diazotrophs, therefore, must protect the particular enzyme system from inactivation by the presence of oxygen in the surrounding environment [88, 91]. As a protective measure, they may permanently live in anaerobic environment, they may develop special O_2 impermeable physical barrier surrounding the nitrogenase, which may reduce the O_2 concentration around the enzyme by special metabolic activity. This will also enhance the O_2 tolerance of the enzyme by structural modification or may balance the inactivated enzyme by increasing its synthesis [91]. These different mechanisms for protecting the enzyme have been reviewed by [91].

5. RECENT METABOLIC/ENZYME ENGINEERING STRATEGIES FOR IMPROVED CG BIOCONVERSION AND H_2 PRODUCTION

Similar to different other products, number of very recent metabolic engineering strategies are known to remarkably improve the H_2 production ability of tested microorganisms [92-95]. Some significant works are listed in Table 1. Proteorhodopsin can absorb light energy and is involved in the translocation of protons through the membranes for ATP synthesis [96]. It can act as a light-driven proton pump and the excess protons generated by this process have the potential to act as substrate for H_2 producing enzymes. Kim et al. (2012) have developed a recombinant *Escherichia coli* strain where proteorhodopsin has been co-expressed with [NiFe]-hydrogenase gene from *Hydrogenovibrio marinus* and nearly 1.3-fold improvement in H_2 production has been reported [96]. Similarly, high pigment content is known to have negative effect on photo-fermentative H_2 production by purple non-sulfur photosynthetic bacteria [97]. In another investigation by [97], pufQ, a key gene involved in bacteriochlorophyll biosynthesis was relocated within the genome of *Rhodobacter capsulatus*. The recombinant strain with relatively lower amount of pigment was able to enhance the H_2 production by 27% [97]. In another study, [FeFe]-hydrogenase gene from *Enterobacter cloacae* IIT BT-08 has been successfully expressed in *E. coli* and improved H_2 production has been reported [100]. Similarly, cyanobacterial bidirectional [NiFe]-hydrogenase has been expressed in *E. coli* and 41 fold increase in hydrogen production has been observed [99]. The genes involved in pyruvate metabolism in *Rhodospirillum rubrum* has been reported to be expressed in recombinant *Rhodobacter sphaeroides*. The recombinant strain was able to increase the H_2 production by nearly two folds [98]. Further, enhanced hydrogen production has also been reported by overexpression of the formate hydrogen lyase (FHL) complex which is responsible for hydrogen production by dissociation of formate to CO_2 and H_2 in case of the organism having PFL pathway [23].

Hydrogen production by glycerol bioconversion was specially targeted in a recent metabolic engineering strategy by [101]. A mutant *E. coli* strain (BW25113 frdC) lacking fumarate reductase activity and less likely towards repression of hydrogen synthesis was chosen. The strain was further improved by adaptive evolution and chemical mutagenesis followed by selection based on improved growth in glycerol containing media. According to the authors, during this process, several pathways responsible for inhibition of H_2 production were repressed and a beneficial pathway was induced resulting in 20 fold increased H_2 yield [101]. In another investigation, glycerol dehydrogenase (GldA) and dihydroxyacetone kinase (DhaKLM) were overexpressed in a recombinant *E. coli* strain and improved Hydrogen production has been reported [102]. Similarly, during glycerol fermentation, H_2 accumulated in the headspace may be recycled as an electron donor for the reduction of fumarate to succinate and it has negative effect on glycerol fermentation. An frdA mutant *E. coli,* devoid of the enzyme fumarate reductase (FRD), was found to be capable of improved glycerol fermentation in presence of H_2 in the headspace [103, 104]. Thus, from the discussion so far it can be concluded that a number of metabolic engineering strategies have been evaluated for hydrogen production. However, only a limited number of such investigations have been undertaken where crude glycerol was considered as the sole substrate for hydrogen production. Therefore, for improved biotransformation of CG to hydrogen, metabolic engineering strategy specially dedicated to entire CG bioconversion pathway is still needed.

Table 1. Recent genetic/ enzyme engineering strategies for improved substrate utilization

Sl no.	Microorganism	Target enzyme /strategy	Manipulated gene /additional strategy	Results	Reference
1	*Recombinant E. coli BL21(DE3)*	Co-expression of proteorhodopsin and [NiFe]-hydrogenase	*Erwinia uredovora* crt E, B, I, Y gene; b-diox gene; pR gene (coding proteorhodopsin) and [NiFe]-hydrogenase gene from *Hydrogenovibrio marinus*	~1.3-fold more H_2 production	[96]
2	*Rhodobacter capsulatus*	To disturbed the enzymes involved in bacteriochlorophyll biosynthesis	Relocation of pufQ (regulatory gene of bacteriochlorophyll biosynthesis)	27% increase in H_2 production	[97]
3	*Rhodobacter sphaeroides*	Combining dark and photo fermentation enzymes	Dark fermentative pyruvate metabolism gene from *Rhodospirillum rubrum*	H_2 production increased up to two fold	[98]
4	*Escherichia coli*	Bidirectional [NiFe]-hydrogenase / inhibition of hydrogen uptake	Hydrogenase gene from the cyanobacterium	41 fold increase in hydrogen production over the wild type	[99]
5	*Escherichia coli*	[FeFe]-hydrogenase	[FeFe]-hydrogenase gene from *Enterobacter cloacae* IIT BT-08	3.12 mol-H_2 yield per mol glucose	[100]
6	*E. coli BW25113 frdC (mutant strain)*	Fumarate reductase (lacking), enolase induced	Further chemical mutagenesis and adaptive evolution leading to repression of the genes of several pathways responsible for decrease hydrogen yields (fbp, focA, and gatYZ) while a beneficial pathway encoding enolase was induced.	Obtained *Escherichia coli* (HW2) that produces 20-fold more hydrogen in glycerol containing medium	[101]
7	*Escherichia coli*	Simultaneous overexpression of glycerol dehydrogenase (gldA) and dihydroxyacetone kinase (dhaKLM)	Glycerol dehydrogenase (gldA), dihydroxyacetone kinase (dhaKLM)	High rate of glycerol utilization	[102]
8	*Escherichia coli*	The *frdA* mutant is devoid of the enzyme fumarate Reductase (FRD), which converts fumarate to succinate	frdA	Improved glycerol fermentation in presence of H_2 in headspace	[103, 104]

CONCLUSION

Biodiesel derived CG is a potential feedstock for fermentative hydrogen production. It is mainly a carbon source; however, it also contains a range of trace elements required for microbial growth. So far, bio-hydrogen production strategies such as, dark fermentation, photo-fermentation and microbial electrolysis cells have been evaluated for CG bioconversion. Owing to the physiological differences between the microorganisms that can be used, more than one enzymatic pathways are available for hydrogen production by CG bioconversion. Further, due to stoichiometric/thermodynamic limitations, maximum possible

H_2 yield per mole of glycerol consumed is different for different pathways. For example, both oxidative and reductive metabolism of glycerol is possible and most microorganisms tested for hydrogen production by CG bioconversion are known to have the enzymes for both the pathways. If glycerol is mostly metabolized through the reductive pathway, most of the hydrogen atoms present in glycerol will be utilized for 1,3-propandiol (as it is a large molecule) production resulting in poor hydrogen yield per mole of glycerol utilized. Choosing, therefore, the right strain or optimizing the process parameters to reduce 1,3-propandiol production might be an interesting strategy for improved hydrogen yield by CG bioconversion. Further, once glycerol is broken down to pyruvate, its further metabolism may proceed via PFOR of PFL pathway and out of these two, PFL pathway does not have the enzyme system required to have the access of NADH for hydrogen production. Hence, maximum possible hydrogen yield for the microorganisms having only PFL pathway is only 2 mole H_2 per mole of glucose. Similarly, during hydrogen production via PFOR or PFL pathway, fermentation end product such as organic acids and solvents are produced and availability of protons for hydrogen production is limited. However, pentose phosphate pathway is another pathway, by successful manipulation of which sugars can be completely converted to H_2O and CO_2 without the formation of any other fermentation end product. Therefore, if glycerol metabolism can be deviated from conventional glycolytic metabolism to pentose phosphate pathway (PPP), maximum hydrogen yield may be possible. In a recent investigation, pure enzymes of PPP were shown to be capable of improving the H_2 yield. However, considering the cost of pure enzymes, if a strategy could be developed where crude preparations of PPP enzymes from different sources is collected and assembled, improved hydrogen production may be possible. Further, Ni, Fe, S and Mo are the different elements essential for the synthesis of different hydrogen producing/consuming enzymes (hydrogenases and nitrogenases). Therefore, proper balance of these elements in the culture media which includes the enzymes may be necessary for enhanced hydrogen yield.

ACKNOWLEDGMENTS

Financial support from Natural Sciences and Engineering Research Council of Canada (Discovery Grant 355254), Ministère des Relations Internationales du Québec (coopération Paraná-Québec 2010-2011), Le Centre de Recherche Industrielle du Québec (CRIQ), MAPAQ (No. 809051) and Fonds Québécois de la Recherche sur la Nature et les Technologies (FQRNT) has been gratefully acknowledged.

REFERENCES

[1] Krupp M, Widmann R. Biohydrogen production by dark fermentation: Experiences of continuous operation in large lab scale. *International Journal of Hydrogen Energy.* 2009;34(10):4509-16.

[2] Rattana S, Junyapoon S, Incharoensakdi A, Phunpruch S. *Hydrogen Production by the Green Alga Scenedesmus sp. KMITL-O1 under Heterotrophic Conditions.*

[3] Ren N, Guo W, Liu B, Cao G, Ding J. Biological hydrogen production by dark fermentation: challenges and prospects towards scaled-up production. *Current Opinion in Biotechnology.* 2011;22(3):365-70.

[4] Kapdan IK, Kargi F, Oztekin R, Argun H. Bio-hydrogen production from acid hydrolyzed wheat starch by photo-fermentation using different Rhodobacter sp. *International Journal of Hydrogen Energy.* 2009;34(5):2201-7.

[5] Guo XM, Trably E, Latrille E, Carrère H, Steyer J-P. Hydrogen production from agricultural waste by dark fermentation: A review. *International Journal of Hydrogen Energy.* 2010;35(19):10660-73.

[6] Sarma SJ, Brar SK, Sydney EB, Le Bihan Y, Buelna G, Soccol CR. Microbial hydrogen production by bioconversion of crude glycerol: A review. *International Journal of Hydrogen Energy.* 2012;37(8):6473-90.

[7] Chatzifragkou A, Papanikolaou S. Effect of impurities in biodiesel-derived waste glycerol on the performance and feasibility of biotechnological processes. *Applied microbiology and biotechnology.* 2012;95(1):13-27.

[8] Santibáñez C, Varnero MT, Bustamante M. Residual glycerol from biodiesel manufacturing, waste or potential source of bioenergy: a review. *Chilean journal of agricultural research.* 2011;71:3.

[9] Koh LP, Ghazoul J. Biofuels, biodiversity, and people: Understanding the conflicts and finding opportunities. *Biological Conservation.* 2008;141(10):2450-60.

[10] Selembo PA, Perez JM, Lloyd WA, Logan BE. Enhanced hydrogen and 1,3-propanediol production from glycerol by fermentation using mixed cultures. *Biotechnology and bioengineering.* 2009;104(6):1098-106.

[11] Liu F, Fang B. Optimization of bio-hydrogen production from biodiesel wastes by *Klebsiella pneumoniae*. *Biotechnology Journal.* 2007;2(3):374-80.

[12] Sabourin-Provost G, Hallenbeck PC. High yield conversion of a crude glycerol fraction from biodiesel production to hydrogen by photofermentation. *Bioresource Technology.* 2009;100(14):3513-7.

[13] Sakai S, Yagishita T. Microbial production of hydrogen and ethanol from glycerol-containing wastes discharged from a biodiesel fuel production plant in a bioelectrochemical reactor with thionine. *Biotechnology and bioengineering.* 2007;98(2):340-8.

[14] Selembo PA, Perez JM, Lloyd WA, Logan BE. High hydrogen production from glycerol or glucose by electrohydrogenesis using microbial electrolysis cells. *International Journal of Hydrogen Energy.* 2009;34(13):5373-81.

[15] Ngo TA, Sim SJ. Dark fermentation of hydrogen from waste glycerol using hyperthermophilic eubacterium *Thermotoga neapolitana*. *Environmental Progress & Sustainable Energy.* 2012;31(3):466-73.

[16] Varrone C, Giussani B, Izzo G, Massini G, Marone A, Signorini A, et al. Statistical optimization of biohydrogen and ethanol production from crude glycerol by microbial mixed culture. *International Journal of Hydrogen Energy.* (0).

[17] Biebl H, Menzel K, Zeng AP, Deckwer WD. Microbial production of 1, 3-propanediol. *Applied microbiology and biotechnology.* 1999;52(3):289-97.

[18] Drożdżyńska A, Leja K, Czaczyk K. Biotechnological production of 1, 3-propanediol from crude glycerol. J Biotechnol, Comput Biol Bionanotechnol. 2011;1:92-100.

[19] Ahrens K, Menzel K, Zeng AP, Deckwer WD. Kinetic, dynamic, and pathway studies of glycerol metabolism by *Klebsiella pneumoniae* in anaerobic continuous culture: III. Enzymes and fluxes of glycerol dissimilation and 1, 3-propanediol formation. *Biotechnology and bioengineering*. 1998;59(5):544-52.

[20] Nemeth A, Kupcsulik B, Sevella B. 1, 3-Propanediol oxidoreductase production with *Klebsiella pneumoniae* DSM2026. *World Journal of Microbiology and Biotechnology*. 2003;19(7):659-63.

[21] Chen X, Sun Y, Xiu ZL, Li X, Zhang D. Stoichiometric analysis of biological hydrogen production by fermentative bacteria. *International Journal of Hydrogen Energy*. 2006;31(4):539-49.

[22] Hallenbeck PC, Abo-Hashesh M, Ghosh D. Strategies for improving biological hydrogen production. *Bioresource Technology*. 2012;110(0):1-9.

[23] Yoshida A, Nishimura T, Kawaguchi H, Inui M, Yukawa H. Enhanced hydrogen production from glucose using ldh-and frd-inactivated *Escherichia coli* strains. *Applied microbiology and biotechnology*. 2006;73(1):67-72.

[24] Kim YM, Cho HS, Jung GY, Park JM. Engineering the pentose phosphate pathway to improve hydrogen yield in recombinant *Escherichia coli*. *Biotechnology and bioengineering*. 2011.

[25] Hallenbeck PC, Ghosh D. Improvements in fermentative biological hydrogen production through metabolic engineering. *Journal of Environmental Management*. 2010.

[26] Zigová J, Šturdík E, Vandák D, Schlosser Š. Butyric acid production by *Clostridium butyricum* with integrated extraction and pertraction. *Process Biochemistry*. 1999;34(8):835-43.

[27] Papanikolaou S, Ruiz-Sanchez P, Pariset B, Blanchard F, Fick M. High production of 1, 3-propanediol from industrial glycerol by a newly isolated *Clostridium butyricum* strain. *Journal of Biotechnology*. 2000;77(2):191-208.

[28] Malaoui H, Marczak R. Purification and characterization of the 1-3-propanediol dehydrogenase of *Clostridium butyricum* E5. *Enzyme and microbial technology*. 2000;27(6):399-405.

[29] Schink B. Energetics of syntrophic cooperation in methanogenic degradation. *Microbiology and Molecular Biology Reviews*. 1997;61(2):262-80.

[30] Herlemann D, Geissinger O, Ikeda-Ohtsubo W, Kunin V, Sun H, Lapidus A, et al. Genomic analysis of "*Elusimicrobium minutum*," the first cultivated representative of the phylum "Elusimicrobia"(formerly termite group 1). *Applied and environmental microbiology*. 2009;75(9):2841-9.

[31] Geissinger O, Herlemann DPR, Mörschel E, Maier UG, Brune A. The ultramicrobacterium "*Elusimicrobium minutum*" gen. nov., sp. nov., the first cultivated representative of the termite group 1 phylum. *Applied and environmental microbiology*. 2009;75(9):2831-40.

[32] Sawers R. Formate and its role in hydrogen production in *Escherichia coli*. *Biochemical Society Transactions*. 2005;33(1):42-6.

[33] Yu M, Zhang Y, Tang I, Yang ST. Metabolic engineering of *Clostridium tyrobutyricum* for n-butanol production. *Metabolic Engineering*. 2011.

[34] Hallenbeck PC, Ghosh D. Improvements in fermentative biological hydrogen production through metabolic engineering. Journal of Environmental Management. 2012;95, *Supplement*(0):S360-S4.

[35] De Vrije T, Mars A, Budde MAW, Lai M, Dijkema C, De Waard P, et al. Glycolytic pathway and hydrogen yield studies of the extreme thermophile *Caldicellulosiruptor saccharolyticus*. *Applied microbiology and biotechnology*. 2007;74(6):1358-67.

[36] Lee KS, Lo YS, Lo YC, Lin PJ, Chang JS. Operation strategies for biohydrogen production with a high-rate anaerobic granular sludge bed bioreactor. *Enzyme and microbial technology*. 2004;35(6):605-12.

[37] Chandrasekaran S, Hotza D. *Bioproduction of Hydrogen with the Assistance of Electrochemical Technology*.

[38] Wamelink M, Struys E, Jakobs C. The biochemistry, metabolism and inherited defects of the pentose phosphate pathway: A review. *Journal of Inherited Metabolic Disease*. 2008;31(6):703-17.

[39] Marin-Valencia I, Cho SK, Rakheja D, Hatanpaa KJ, Kapur P, Mashimo T, et al. Glucose metabolism via the pentose phosphate pathway, glycolysis and Krebs cycle in an orthotopic mouse model of human brain tumors. *NMR in Biomedicine*. 2012:n/a-n/a.

[40] Jones NP, Schulze A. Targeting cancer metabolism – aiming at a tumour's sweet-spot. *Drug Discovery Today*. 2012;17(5–6):232-41.

[41] Berg JM, Tymoczko JL, Stryer L. *The Metabolism of Glucose 6-Phosphate by the Pentose Phosphate Pathway Is Coordinated with Glycolysis*. 2002.

[42] Cairns RA, Harris IS, Mak TW. Regulation of cancer cell metabolism. *Nature Reviews Cancer*. 2011;11(2):85-95.

[43] Tong X, Zhao F, Thompson CB. The molecular determinants of de novo nucleotide biosynthesis in cancer cells. *Current opinion in genetics & development*. 2009;19(1):32-7.

[44] Heyer NI, Woodward J. Efficient hydrogen production using enzymes of the pentose phosphate cycle. Proceedings of the 2001 DOE Hydrogen Program Review, NREL/CP-570-30535 *Accessed February*. 2001;2:2010.

[45] Woodward J, Orr M, Cordray K, Greenbaum E. Biotechnology: Enzymatic production of biohydrogen. *Nature*. 2000;405(6790):1014-5.

[46] Mertens R, Liese A. Biotechnological applications of hydrogenases. *Current Opinion in Biotechnology*. 2004;15(4):343-8.

[47] Kim DH, Kim MS. Hydrogenases for biological hydrogen production. *Bioresource Technology*. 2011.

[48] Evans DJ, Pickett CJ. Chemistry and the hydrogenases. *Chem Soc Rev*. 2003;32(5):268-75.

[49] Vignais P, Colbeau A. Molecular biology of microbial hydrogenases. *Current issues in molecular biology*. 2004;6(2):159-88.

[50] Kleihues L, Lenz O, Bernhard M, Buhrke T, Friedrich B. The H_2 sensor of *Ralstonia eutropha* is a member of the subclass of regulatory [NiFe] hydrogenases. *Journal of bacteriology*. 2000;182(10):2716-24.

[51] Lenz O, Bernhard M, Buhrke T, Schwartz E, Friedrich B. The hydrogen-sensing apparatus in *Ralstonia eutropha*. *Journal of molecular microbiology and biotechnology*. 2002;4(3):255-62.

[52] Veit A, Akhtar MK, Mizutani T, Jones PR. Constructing and testing the thermodynamic limits of synthetic NAD(P)H:H2 pathways. *Microbial Biotechnology.* 2008;1(5):382-94.
[53] Meyer J. [FeFe] hydrogenases and their evolution: a genomic perspective. *Cellular and molecular life sciences.* 2007;64(9):1063-84.
[54] Thauer RK, Klein AR, Hartmann GC. Reactions with molecular hydrogen in microorganisms: evidence for a purely organic hydrogenation catalyst. *Chemical reviews.* 1996;96(7):3031.
[55] Buurman G, Shima S, Thauer RK. The metal-free hydrogenase from methanogenic archaea: evidence for a bound cofactor. *FEBS Letters.* 2000;485(2–3):200-4.
[56] Calusinska M, Happe T, Joris B, Wilmotte A. The surprising diversity of clostridial hydrogenases: a comparative genomic perspective. *Microbiology.* 2010;156(6):1575-88.
[57] Tamagnini P, Leitao E, Oliveira P, Ferreira D, Pinto F, Harris DJ, et al. Cyanobacterial hydrogenases: diversity, regulation and applications. *FEMS Microbiology Reviews.* 2007;31(6):692-720.
[58] Thauer RK, Kaster A-K, Goenrich M, Schick M, Hiromoto T, Shima S. Hydrogenases from methanogenic archaea, nickel, a novel cofactor, and H_2 storage. *Annual review of biochemistry.* 2010;79(1):507-36.
[59] Rákhely G, Kovács ÁT, Maróti G, Fodor BD, Csanádi G, Latinovics D, et al. Cyanobacterial-type, heteropentameric, NAD^+-reducing NiFe hydrogenase in the purple sulfur photosynthetic bacterium *Thiocapsa roseopersicina. Applied and environmental microbiology.* 2004;70(2):722-8.
[60] Pinske C, Sawers RG. Delivery of Iron-sulfur clusters to the hydrogen-oxidizing [NiFe]-hydrogenases in *Escherichia coli* requires the A-type carrier proteins ErpA and IscA. *PLoS ONE.* 2012;7(2):e31755.
[61] Adams MWW. Biochemical diversity among sulfur-dependent, hyperthermophilic microorganisms. *FEMS Microbiology Reviews.* 1994;15(2-3):261-77.
[62] Albracht SPJ. Nickel hydrogenases: in search for the active site. *Biochimica et Biophysica Acta G General Subjects.* 1994;1188:167-204.
[63] Friedrich B, Schwartz E. Molecular biology of hydrogen utilization in aerobic chemolithotrophs. *Annual Reviews in Microbiology.* 1993;47(1):351-83.
[64] Happe RP, Roseboom W, Plerik A, Albracht SPJ, Bagley KA. Biological activation of hydrogen. *Nature.* 1997;385(6612):126.
[65] Ogata H, Hirota S, Nakahara A, Komori H, Shibata N, Kato T, et al. Activation process of [NiFe] hydrogenase elucidated by high-resolution x-ray analyses: conversion of the ready to the unready state. *Structure* (London, England : 1993). 2005;13(11):1635-42.
[66] Pandelia ME, Ogata H, Currell LJ, Flores M, Lubitz W. Inhibition of the [NiFe] hydrogenase from Desulfovibrio vulgaris Miyazaki F by carbon monoxide: An FTIR and EPR spectroscopic study. *Biochimica et Biophysica Acta (BBA)-Bioenergetics.* 2010;1797(2):304-13.
[67] Cammack R. *Hydrogen as a fuel: learning from nature:* CRC; 2001.
[68] Ogata H, Mizoguchi Y, Mizuno N, Miki K, Adachi S, Yasuoka N, et al. Structural studies of the carbon monoxide complex of [NiFe] hydrogenase from *Desulfovibrio vulgaris* Miyazaki F: suggestion for the initial activation site for dihydrogen. *Journal of the American Chemical Society.* 2002;124(39):11628-35.

[69] Matsumoto T, Kabe R, Nonaka K, Ando T, Yoon KS, Nakai H, et al. Model study of CO inhibition of [NiFe] hydrogenase. *Inorganic chemistry.* 2011.
[70] Smith LA, Hill S, Yates M. *Inhibition by acetylene of conventional hydrogenase in nitrogen-fixing bacteria.* 1976.
[71] Sun JH, Hyman MR, Arp DJ. Acetylene inhibition of *Azotobacter vinelandii* hydrogenase: acetylene binds tightly to the large subunit. Biochemistry. 1992;31(12):3158-65.
[72] Maness PC, Yu J, Eckert C, Ghirardi ML. Photobiological hydrogen production–prospects and challenges. *Microbe.* 2009;4:275-80.
[73] Cornish AJ, Gärtner K, Yang H, Peters JW, Hegg EL. Mechanism of proton transfer in [FeFe]-hydrogenase from *Clostridium pasteurianum. Journal of Biological Chemistry.* 2011;286(44):38341-7.
[74] Czech I, Stripp S, Sanganas O, Leidel N, Happe T, Haumann M. The [FeFe]-hydrogenase maturation protein HydF contains a H-cluster like [4Fe4S]–2Fe site. *FEBS Letters.* 2011;585(1):225-30.
[75] Rudolf K. Thauer ES, Hamilton WA, Barton LL. *Energy metabolism and phylogenetic diversity of sulphate-reducing bacteria Sulphate-Reducing Bacteria:* Cambridge University Press; 2007.
[76] Mulder DW. *Biochemical, spectroscopic, and structural investigations on [FeFe]-hydrogenase maturation and complex metallocluster assembly:* Montana State University; 2010.
[77] Kamp C, Silakov A, Winkler M, Reijerse EJ, Lubitz W, Happe T. Isolation and first EPR characterization of the [FeFe]-hydrogenases from green algae. *Biochimica et Biophysica Acta (BBA)-Bioenergetics.* 2008;1777(5):410-6.
[78] Horner DS, Heil B, Happe T, Embley TM. Iron hydrogenases-ancient enzymes in modern eukaryotes. *Trends in biochemical sciences.* 2002;27(3):148-53.
[79] Adams M. The structure and mechanism of iron-hydrogenases. *Biochimica et biophysica acta.* 1990;1020(2):115.
[80] Barstow B, Agapakis C, Boyle P, Grandl G, Silver P, Wintermute E. A synthetic system links FeFe-hydrogenases to essential *E. coli* sulfur metabolism. Journal of Biological Engineering. 2011;5(1):7.
[81] Frey M. *Hydrogenases: hydrogen-activating enzymes. ChemBioChem.* 2002;3(2-3):153-60.
[82] Shima S, Thauer RK. A third type of hydrogenase catalyzing H_2 activation. *The Chemical Record.* 2007;7(1):37-46.
[83] Vignais PM, Billoud B. Occurrence, classification, and biological function of hydrogenases: an overview. *ChemInform.* 2007;38(50):no-no.
[84] Thauer RK. Biochemistry of methanogenesis: a tribute to Marjory Stephenson. *Microbiology.* 1998;144(9):2377-406.
[85] Rubio LM, Ludden PW. Biosynthesis of the iron-molybdenum cofactor of nitrogenase. *Annual Review of Microbiology.* 2008;62(1):93-111.
[86] Eddin HS. *Investigation of molybdenum iron protein expression and activity in Wolinella succinogenes* 2010.
[87] Raymond J, Siefert JL, Staples CR, Blankenship RE. The natural history of nitrogen fixation. *Molecular Biology and Evolution.* 2004;21(3):541-54.

[88] Tamagnini P, Axelsson R, Lindberg P, Oxelfelt F, Wünschiers R, Lindblad P. Hydrogenases and hydrogen metabolism of cyanobacteria. *Microbiology and Molecular Biology Reviews.* 2002;66(1):1-20.

[89] McKinlay JB, Harwood CS. Photobiological production of hydrogen gas as a biofuel. *Current Opinion in Biotechnology.* 2010;21(3):244-51.

[90] Drepper T, Groß S, Yakunin AF, Hallenbeck PC, Masepohl B, Klipp W. Role of GlnB and GlnK in ammonium control of both nitrogenase systems in the phototrophic bacterium *Rhodobacter capsulatus. Microbiology.* 2003;149(8):2203-12.

[91] Soto-Urzúa L, Baca B. Mechanisms for protecting nitrogenase from inactivation by oxygen. *Revista latinoamericana de microbiología.*43(1):37.

[92] Lee SY, Lee HJ, Park J-M, Lee JH, Park J-S, Shin HS, et al. Bacterial hydrogen production in recombinant *Escherichia coli* harboring a HupSL hydrogenase isolated from *Rhodobacter sphaeroides* under anaerobic dark culture. *International Journal of Hydrogen Energy.* 2010;35(3):1112-6.

[93] Eltsova Z, Vasilieva L, Tsygankov A. Hydrogen production by recombinant strains of *Rhodobacter sphaeroides* using a modified photosynthetic apparatus. *Applied Biochemistry and Microbiology.* 2010;46(5):487-91.

[94] Cho H-S, Kim YM, Min BE, Jung GY, Park JM. Improvement of hydrogen production yield by rebalancing NADPH/NADH ratio in a recombinant *Escherichia coli. Journal of Nanoelectronics and Optoelectronics.* 2011;6(3):343-7.

[95] Seol E, Manimaran A, Jang Y, Kim S, Oh Y-K, Park S. Sustained hydrogen production from formate using immobilized recombinant *Escherichia coli* SH5. *International Journal of Hydrogen Energy.* 2011;36(14):8681-6.

[96] Kim J, Jo B, Jo Y, Cha H. Improved production of biohydrogen in light-powered *Escherichia coli* by co-expression of proteorhodopsin and heterologous hydrogenase. *Microbial Cell Factories.* 2012;11(1):1-7.

[97] Ma C, Wang X, Guo L, Wu X, Yang H. Enhanced photo-fermentative hydrogen production by *Rhodobacter capsulatus* with pigment content manipulation. *Bioresource Technology.* 2012;118(0):490-5.

[98] Kim E-J, Kim M-S, Lee JK. Hydrogen evolution under photoheterotrophic and dark fermentative conditions by recombinant *Rhodobacter sphaeroides* containing the genes for fermentative pyruvate metabolism of *Rhodospirillum rubrum. International Journal of Hydrogen Energy.* 2008;33(19):5131-6.

[99] Maeda T, Vardar G, Self W, Wood T. Inhibition of hydrogen uptake in *Escherichia coli* by expressing the hydrogenase from the cyanobacterium Synechocystis sp. *PCC 6803. BMC Biotechnology.* 2007;7(1):25.

[100] Chittibabu G, Nath K, Das D. Feasibility studies on the fermentative hydrogen production by recombinant *Escherichia coli* BL-21. *Process Biochemistry.* 2006;41(3):682-8.

[101] Hu H, Wood TK. An evolved *Escherichia coli* strain for producing hydrogen and ethanol from glycerol. *Biochemical and Biophysical Research Communications.* 2010;391(1):1033-8.

[102] Durnin G, Clomburg J, Yeates Z, Alvarez PJJ, Zygourakis K, Campbell P, et al. Understanding and harnessing the microaerobic metabolism of glycerol in *Escherichia coli. Biotechnology and bioengineering.* 2009;103(1):148-61.

[103] Murarka A, Dharmadi Y, Yazdani SS, Gonzalez R. Fermentative utilization of glycerol by *Escherichia coli* and its implications for the production of fuels and chemicals. *Applied and environmental microbiology.* 2008;74(4):1124-35.

[104] Shams Yazdani S, Gonzalez R. Engineering *Escherichia coli* for the efficient conversion of glycerol to ethanol and co-products. *Metabolic Engineering.* 2008;10(6):340-51.

In: Enzymes in Value-Addition of Wastes
Editors: Satinder Kaur Brar and Mausam Verma
ISBN: 978-1-62948-968-1

Chapter 11

ROLE OF ENZYMES IN FROZEN DOUGH BAKERY PRODUCTS

Ali Asghar[1], Saeed Akhtar[2] and Muhammad Nadeem Akhtar[3]
[1]National Institute of Food Science and Technology, University of Agriculture, Faisalabad, Pakistan
[2]Department of Food Science and Technology, University College of Agriculture, Bahauddin Zakariya University, Multan, Pakistan
[3]School of Food Sciences, The University of Faisalabad, Faisalabad, Pakistan

ABSTRACT

The quality of bakery products could be improved by the use of a variety of additives, but the use of natural additives is gaining popularity. Value addition for raw commodities is the slogan of modern agriculture. Frozen dough products are one recent example of value-added cereal products. Water distribution changes that usually occur during extended frozen storage and freeze-thaw cycles contribute to the extended proofing time and reduced loaf volume of frozen bread dough, and thus result in stability problems in final products. Enzymes are also considered potential natural additives that have the ability to interact or modify the starches and gluten network in a dough system, and thus behave as dough improvers and strengtheners.

Keywords: Frozen dough, enzymes, bread quality

INTRODUCTION

There was a time when the production and export of only raw commodities was considered a priority achievement for a developing economy. However, mechanization and adding value to raw commodities has become the emphasis in the modern era. With increasing urbanization and industrialization, there is rapid development towards the value addition sector in every industry. Among food processing industries, frozen and refrigerated dough bakery products are considered among one of the fastest growing segments.

Baking is one of the oldest crafts in the world. Nevertheless, the majority of baked foodstuffs have a little shelf-life as a result of staling. Therefore, frozen dough technology has been developed in the early twentieth century (Sluimer, 2005). The technique gives customer benefits by allowing the baker or retailer to supply fresh bread at almost any time. Consequently, this technology has become one of the most significant technologies practiced by today's baking industry. Numerous studies relating to frozen dough have been available, but much of the information needed to produce premium frozen dough has been kept as proprietary technology and is not publicly available. This has slowed down the growth of better and improved processes. Presently, frozen dough is being used throughout the world, resulting in economic benefits to the producer and increasing ease for the user. Besides many advantages, frozen dough also shows several problems in the quality of the finished products.

With every passing day, the use of new additives is becoming a common practice in the business of baking. Due to rapid increase in the frozen dough production and utilization, problems arise that can be solved by the use of certain additives and chemical agents. All the compounds that interact with water can affect the quality of the resulting bakery product. A lot of compounds and additives are available that have water binding and gelling properties. A wide range of additives are available and natural additives are needed to meet current consumer concepts of good nutrition.

The main factor affecting frozen dough is the yeast's stability after freezing. Hence, yeast problems have been extensively studied (Ribotta et al., 2003). Throughout the dough processing, while the temperature comes near the freezing point, ice crystals damage the yeast cell walls. As a result of this, the yeasts release glutathione, a protein reducing agent, which accelerate the weakening of the dough. To avoid these effects, oxidants, such as potassium bromate (KBrO3) and ascorbic acid are added into yeast-leavened dough. Ascorbic acid is now frequently used in baked products, particularly bread. Potassium bromate is a physically powerful oxidant that was also once used, although its use has declined in recent years.

Many food additives for example, enzymes and amino acids have been used in breadmaking as bromate replacements (Morita et al., 1997). After usage of $KBrO_3$ was forbidden in numerous countries, interest developed in searching for an enzyme to replace this chemical oxidant (Mathewson, 1998). The benefit of enzyme usage has been demonstrated in breadmaking. A lot of studies (Ribotta and Le Bail, 2007; Guy, 2001;Hille and Schooneveld-Bergmans, 2004) reveal that enzymes play crucial roles in bread making such as increasing loaf volume, producing finer crumb cells, and extending shelf-life. Furthermore, Japanese Patent specification No. 5701/1968 reveals a method for bread quality enhancement by adding a component to the dough which contained hemicellulase. The Patent claims "the combination of the enzyme preparation of the invention and lecithin can advantageously replace bromate conventionally used as a baking additive" (U.S. Patent No. 4,990,343., 1991). Ribotta and Le Bail (2007) studied the combination of ascorbic acid and hemicellulase was used to study the thermo-mechanical behavior of dough systems during research. They found that frozen wheat dough consists of an ice phase as well as an unfrozen phase with a very low freezing point. The low freezing point is attributed to freeze concentration of solutes in the unfrozen phase when water is removed as ice crystals during the freezing process. Within this unfrozen phase, deteriorations are taking place even at very low temperatures. They investigated the effects of α-amylase, ascorbic acid, protease, hemicellulase, gluten and guar gum on the thermo-mechanical properties of dough at sub-zero temperature.

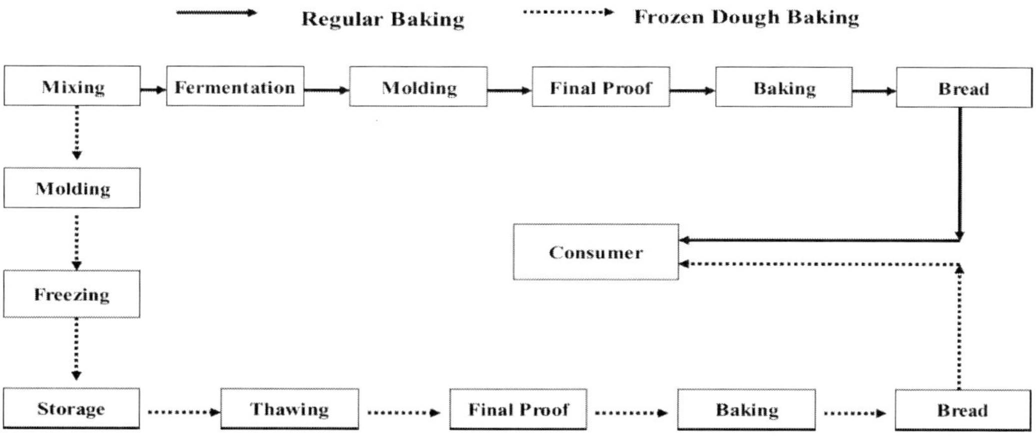

Dotted lines represent the frozen dough processing line. Adapted from Giannou et al., 2006.

Figure 1. Process flow diagram of normal and frozen dough baking. — Continuous line is the ordinary dough production line.

HISTORY, EVOLUTION AND POTENTIAL OF FROZEN DOUGH PRODUCTS

Although bread making is one of the oldest technologies of human history, yet this technology continues to develop and evolve. Even the people of prehistoric time were aware of making breads to fulfill the basic human nutritional needs. If we look into history, we find that between 4000 and 3500 B.C. Egyptians were the first to use yeast as the leavening agents in the baking process. Since its inception, the baking industry has undergone evolutionary steps such that storage of dough under freezing conditions is now a common practice that results in ease in handling. transportation and production of "fresh" bread. Freezing technology is commonly being employed for the preservation of food commodities. In the last decade, production of frozen dough has increased greatly because of direct sales to the consumers and the growing number of in-store bakeries.

The frozen dough product segment is the third largest in the baking industry. Demand for quick and convenient food items is increasing due to changing life styles. The fast-developing frozen dough industry is providing easy alternatives to traditional bakery products. Consumers who make bread from frozen dough expect products with satisfactory quality and sensory characteristics that should not differ much from the traditional fresh ones.

Frozen dough is one of the emerging markets in the world. Due to high market potential and current lack of competition, there is a huge potential of growth of this sector. The total market for frozen dough products covers retail grocery sales and food service (Matuda et al., 2005; Giannou et al., 2006). Now, due to increasing consumer demand for the frozen dough products, several major industrial groups are entering this business (Holcomb, 2000). Increasing growth in the frozen dough business observed in the recent years accompanies the trend for more meals prepared outside of the home (Lou and Wilson, 1998).

Problems with Frozen Dough Products

Water distribution changes that occur during extended frozen storage and freeze thaw cycles contribute to the extended proofing time and reduced loaf volume of frozen dough breads. Ice crystal formation is also a major cause of weakening of the protein network in the dough system that is responsible for gas retention (Cook, 1994; Schroeder, 1999). Freezing and thawing in the frozen dough leads to decreasing the dough strength which also results in the reduction of final loaf volume of breads (Berglund et al., 1991; Varriano et al., 1980; Bloksma, 1990; Resanen et al., 1997; Inoue and Bushuk, 1992 and Holmes and Hoseney, 1987).

Shelf-life of fresh bakery products is usually a few days and after proper baking, the product, i.e. bread, is usually of good quality and sensory features. Fresh bread usually presents an appealing brownish and crunchy crust, a pleasant roasting aroma, fine slicing characteristics, a soft and elastic crumb texture, and a moist mouth feel. During the storage of fresh bread and other bakery items a number of chemical and physical alterations occur, known as staling. As a result of these changes, bread quality deteriorates gradually as it loses its freshness and crispiness while crumb firmness and rigidity increase.

The lower gas retention in frozen dough is mainly due to its high breaking force when thawed. During the cooling stage, consistent changes in the moisture content occur that contribute to a temperature gradient within the bread. Usually the interior of the bread has more moisture than its outer layers. This produces a difference in vapor pressure between the crust and crumb, resulting in moisture migration during the cooling stage. Part of the moisture travels from the crust to the surrounding atmosphere. This mass transfer resistance that exists at the interface between the bread crust and the atmosphere around it influences the rate of moisture migration (Yousif, 1998; Asghar et al., 2005)

In order to control these stability problems in frozen dough, new processing techniques and additives that increase shelf life, enhance quality, and retain stability; sensory and nutritional characteristics of the frozen dough bakery products during prolonged periods of storage are recommended.

Enzymes in Bread and Frozen Dough Baking

Enzymes have a wide range of application in different industrial processing and in baking α-amylase and protease are widely used. Staling is the major reason of spoilage in bakery products. Because stale products become less acceptable to consumers, huge economic losses to the baking industry result (Turgeon et al., 1992; Bechtel et al., 1953; Poutanen, 1997). Different enzymes are currently added to the bread making process for improving dough handling, fresh bread quality and shelf life (Haros et al., 2002). Several other enzymes also act as dough improvers by modifying one of the major dough components (Maarel, 2002).

During the process of staling, different physicochemical transformations occur. The most important are the retrogradation of starch, gluten–starch interactions, and moisture redistribution between and among components. The most important change associated with bread staling is the gradual increase in crumb firmness (Gray and Bemiller, 2003; Schultz et al., 1952). Enzymes are commonly used to reduce the rate of staling in bakery products.

Enzymes are being used increasingly to improve the product quality and functional properties of the frozen dough bakery items.

YEASTS

The main purpose of yeast in the yeast-raised dough is leavening. During dough fermentation, yeast transfer sugars to carbon dioxide (CO2) and alcohol. Also, dough temperature is raised by fermenting. Meanwhile, fermentation flavours are developed (Rogers, 2004).

Baker's yeast has been grouped into two common types depending upon its stability and how it is processed. The first type, fresh yeasts can be compressed, crumbled, or cream types. The other type, dry yeasts, can be active, dry or instant dry yeast. The type of yeast that is suitable for frozen dough making has been to some extent controverted over time. A study conducted by Wolt and D'appolonia (1984) revealed that fresh compressed yeast usually performed better than active dry yeast in proof-time stability over a storage period. Conversely, El-Hady and his coworkers (1996) described that active dry yeast is of better-quality than compressed yeast in maintaining shelf life in frozen dough. This yeast was isolated from banana peel and identified as *Kluyveromyces thermotolerans and Saccharomyces cerevisiae*. The new freeze-tolerant yeasts are claimed to provide good quality bread similar to that made from unfrozen dough (Hino et al., 1987).

HEMICELLULASES

Hemicellulase is a hydrolytic enzyme which can hydrolyze the hemicellulose present in plant cell walls. Hemicellulose is a minor gum-like fraction (~3%) in white flour. In whole wheat flour, the total content ranges from 4 to 7% (Hille and Schooneveld-Bergmans, 2004).

The main use of hemicellulase is in the baking industry. It is used in baked goods, cake mixes and frozen dough. The enzyme increases the quality of the dough and helps with storage life.

Alcoholic beverages and fruit juices can be produced using this type of enzyme. In particular, hemicellulases are used in the production of wines. They help eliminate unnecessary compounds from the skins of the grapes that might contaminate the taste of the wine.

Though plants produce hemicellulases for growth and development, the majority of the commercial interest is in the enzymes produced by microorganisms. The enzymes are produced commercially from fungi and bacteria. In some cases, these organisms have been genetically engineered to prepare optimal quantity of hemicellulase.

The significant function of hemicellulases in many biotechnological applications has led to a great deal of interest in determining their structure and function. The enzymes are often in a modular form; different parts of their structures have varying functions. For this type of enzyme, a part is specialized to bind to the hemicellulose, while another part breaks down the bond connecting the sugars by a process known as hydrolysis This involves adding a molecule of water to the bond, which causes it to break open.

Microorganisms that produce hemicellulases are also found in the digestive tract of humans and higher animals. Since humans cannot degrade hemicellulose, it is considered fiber. The microorganisms in the human body break down some of the hemicellulose, helping it to be digested. Supplements of hemicellulase are available, but as with any supplement, one should consult with a doctor before consuming them.

APPLICATION OF HEMICELLULASE TO BAKED PRODUCTS

The book "Enzymes" (Mathewson, 1998) contains a description of the dough improvement/oxidizing functions of enzyme usage in bread making. It elaborates the fact that more and more researchers have tried to find a single enzyme to act as a bromate replacement but that "no single enzyme has been identified that can replace the addition of oxidizing agents. A number of companies have introduced enzyme preparations containing multiple forms of enzyme activity, including amylase, proteases, and several hemicellulases in combination with ascorbic acid."

Hille and Schooneveld-Bergmans (2004) concluded that both fungal and bacterial hemicellulases are able to enhance fresh bread quality as considered by loaf size, shape, crumb texture and softness. Hammond (1994) concluded that hemicellulase enlarged the loaf volume of pan bread, together with an improvement of shape and symmetry, in addition to finer cell structure and a resilient loaf. Hemicellulases also contributed finer cell structure, whiter crumb, and significantly increased softness in French baguettes. When a mixture of hemicellulase and fungal alpha-amylase was added into white and whole meal loaf formulae, the results illustrated considerably greater volume than that of fungal alpha-amylase alone (Hammond, 1994). Guy (2001) showed that "hemicellulases have been claimed to improve on the effects achieved with fungal amylase and to also improve the fineness of the cellular structure of the crumb. Hemicellulose particles may upset the gluten network. When endoxylanase degrades the WU-AX (water-unextractablearabinoxylan) polymer into a WE-AX (water-extractable arabinoxylan), the gluten net can have better gas retention, resulting in extra volume (DSM 2005).

XYLANS

Xylan is the main type of hemicellulose and its hydrolysis depends on two classes of enzymes. The Endoxylanases (EC 3.2.1.8) cleave the xylan backbone into smaller parts like oligosaccharides that are further degraded to xylose by xylosidases (EC 3.2.1.37). During the last few decades, hugeconcentration has been developed in xylan and its hydrolytic enzymatic complex, for application in bread production, supplement in animal feed, preparation of drinks, textiles, bleaching of cellulose pulp, ethanol and xylitol production.

In cereals, arabinoxylans are among the major non-starch polysaccharides. They constitute 4-8% of barley kernel and represent 25 and 70% of the cell wall polysaccharides of endosperm and aleurone layer, respectively. The arabinoxylans are partially water-soluble and result in highly viscous aqueous solution (Collins et al., 2005). Cereal xylans contain hugeamounts of L-arabinose and consequently, often referred as arabinoxylanswhile, hardwood xylans are often referred to as glucuronoxylans due to large amount of D-glucoronic acid attached to the backbone. Arabinose is linked to the backbone of xylanvia α-1, 2 or α-1, 3 linkage either as single residues or as short side chains. The side chains also contain xylose α- 1, 2-linked to arabinose, and galactose, which can be either β-1, 5-linked to arabinose or α-1,4- linked to xylose.

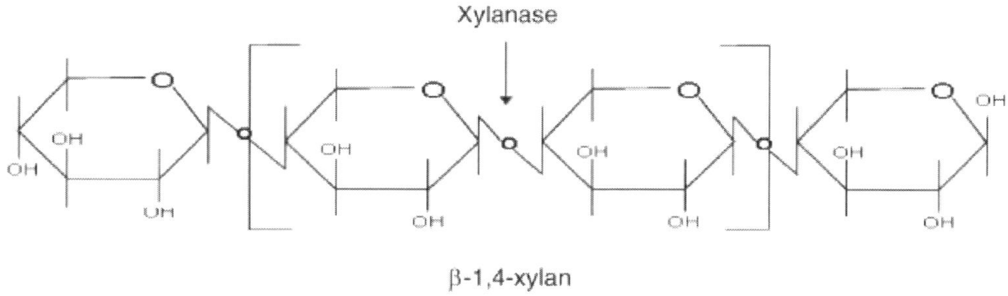

Figure 2. Structure of xylan, xylopyranosyl residues linked through b-1,4 linkages; point of xylanase action is shown.

Bread is the most commonly used and traditional food product around the world. For years, enzymes, such as malt and fungal α-amylases are being used in bread making process. Due to the advancements in baking industry and the ever-increasing demand for more natural products, enzymes have gained much importance in baking industry. They improve dough characteristics and bread quality leading to improved dough flexibility, machinability, stability, loaf volume and crumb structure (Baillet, 2003; Guy and Sarabjit, 2003). Many enzymes, such as proteases, xylanase and cellulases improve the strength of the gluten network and therefore, improve the quality of bakery products (Gray and BeMiller, 2003).

Xylanases are well-known dough conditioners and have reportedly been used to increase loaf volume and the addition of these enzymes resulted in a greater firming rate (Jiang et al., 2005; Armero and Collar, 1998; Bollain et al., 2005) and this positive effect on bread volume is due to the redistribution of water from the pentosan phase to the gluten phase. Xylanases are also known to have anti-staling action during bread storage (Haros et al., 2002).

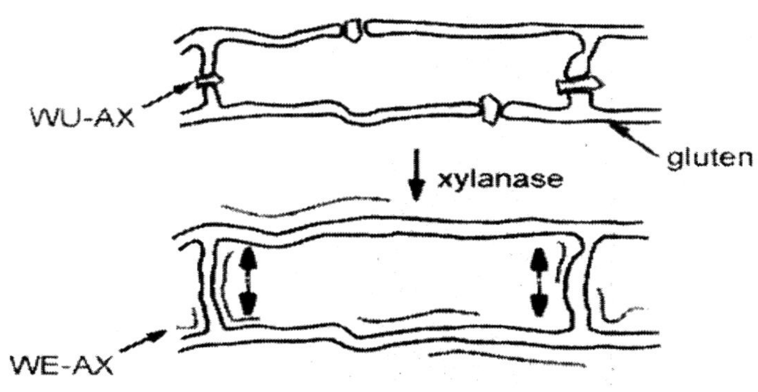

Figure 3. Interactions of arabinoxylan and gluten in dough. WU-AX is water-unextractable arabinoxylans. WE-AX is water-extractable arabinoxylans.

Xylanases are identified to have an anti-staling action during bread storage though the mechanism is not clear. Their improving action might be attributed to the hydrolysis of the cell wall polysaccharides of the wheat grain. The simpler carbohydrates (monosaccharides and oligosaccharides) resulting from the enzyme action could affect the water balance and may interfere with protein-starch interaction during bread storage, in the same way as specific dextrins have been described to interfere in the amylopectin retrogradation (Haros et al., 2002).

LIPASE(S)

Lipases are extensively distributed in nature, plants, animals and microorganisms. Lipases are the enzymes which hydrolyze insoluble fats and fatty acid esters occurring in separate, non-aqueous phases (Underkofler, 1972). According to Hoseney (1998), "all cereals have lipase activity, but the activity varies extensively between cereals, with oats and pearl millet having relatively high activity compared to that of wheat or barley." When different techniques or different substrates are used, it is very difficult to compare the lipases from different cereals.

Lipase hydrolyzes a triglyceride into mono- and diglycerides. The reaction products function as dough emulsifiers and their improving action is based on this mechanism (Sluimer, 2005). Even though lipase is not commonly used in baked foods, it has been widely known to benefit bread quality (Sluimer, 2005, Sahi and Guy, 2004). For example, Gélinas et al. (1998) concluded that lipase has dough bleaching activity if combined with peroxidase and linoleic acid. Moreover, the combinations of lipase, oxidized oil and linoleic acid significantly degraded flour pigments (Mercier and Gélinas, 2001). On the other hand, Underkofler (1972) showed that "lipase activity in flour for baking is objectionable because free fatty acids have a harmful effect in doughs."

Lipases can improve crumb softness of bread. Addition of specific lipases in combination with triglycerides also improves loaf volume, crumb softness, staling rate and flavor. Antistaling functions of emulsifiers are usually due to their interaction with amylopectin that prevents amylopectin crystallization (Hamer, 1995; Lagendijk and Pennings, 1970). The mechanism of monoglycerides in retarding the firming process is based on the ability of monoglycerides to form complexes with amylose (Stampfli and Nersten, 1995; Valjakka et al., 1994).

TRANSGLUTAMINASE

Transglutaminase (protein-glutamine-glutamyltransferase, EC 2.3.2.13) catalyses the acyl-transferase reaction between the carboxymidic group of peptides which contains glutamine residues (acyl donors), and various primary amines (acyl acceptors) including the amino groups of lysine bound in proteins. In the absence of substrates bearing amino groups transglutaminase catalyzes the deamination of glutamine residues in the course of which water molecules are used as acyl acceptors. Transglutaminase can modify proteins by the incorporation of amine, by cross-linking, and by deamination. These enzyme-catalyzed reactions may be used for the modification of food proteins which can lead, for example, to the formation of texture as well as to the protective effect on lysine present in food proteins and thus to a higher retention of the nutritional value.

Transglutaminases are widely distributed in animal tissues and body fluids, plants, fishes and microorganisms. The application of transglutaminase in baking bread and croissants has a positive effect on sensory and nutritional properties. The addition of transglutaminase (5000 ppm/weight of flour) revealed a positive effect on the texture and the distribution of pores which resulted in higher loaves of bread and pastry. The best results in dough raising were achieved by the addition of transglutaminase to the dough which was subsequently frozen (the activation of the enzyme during freezing). The positive effect of the addition of transglutaminase persisted even during the storage of frozen croissants for 90 days with the bread crumb being better than that in the control. The study conducted by Hozová et al. (2002) shows improved concentration of lysine content by the addition of transglutaminases in the frozen dough as shown below.

Transglutaminases are widely distributed in animal tissues and body fluids plants, fishes and microorganisms. The addition of transglutaminase (5000 ppm/weight of flour) revealed a positive effect on the texture and the distribution of pores which resulted in higher loaves of bread and pastry. The best results in dough raising were achieved by the addition of transglutaminase to the dough which was subsequently frozen (the activation of the enzyme during freezing). The positive effect of the addition of transglutaminase persisted even during the storage of frozen croissants for 90 days with the bread crumb being better than that in the control.

PROTEASE

Protease that is commonly present in flour due to the microbial contamination plays a major role in gluten disaggregation during the mixing process. Sometimes, protease is added to high protein hard wheat flour to hydrolyze gluten. This process, termed the mellowing of the gluten, also helps to improve the mixing properties of the dough (Van Dam and Hille, 1992). The function of the protease is to depolymerize and weaken the gluten structure of the dough. The function of protease is to some extent contrary to the action of oxidants and thus helps to control bread texture and improve flavor (Turgeon et al., 1992; Hamer, 1995; Mathewson, 2000; El-Dash and Johnson, 1967). As proteases modify gluten protein, these interactions are weakened and firming is decreased. Proteases also affect the sensory attributes and texture (El-Dash and Johnson, 1967). Sometimes enzymes also show synergistic effects for dough conditioning and for extending shelf -life (Qi Si, 1997; Sato et al., 1991; Sato et al., 1995; Collar et al., 2000). The synergistic combination of xylanases with amylase has proven to be beneficial and is used in many improved formulations to control bread staling (Haseborg and Himmelstein, 1998; Rouau et al., 1994; Maat et al., 1994).

Conclusion and Future Trends

The baking industry has been exploiting the advantages and applications of freezing technology in several frozen dough bakery applications from the last few years. There is an increase in demand for bakery products with increased shelf life from *both* ordinary consumers and food service industries. With the passage of time there is continual development of new frozen dough and other bakery products. One such recent trend is the use of certain natural additives including different enzymes for a wide range of applications due to their ability to bind the excess moisture which prevent its availability to fungus and other microflora that are major causes of deterioration and staling in the bakery products. Sometimes enzymes also show synergistic effects for dough conditioning and for extending shelf –life.

References

Armero, E. & Collar, C. (1998). Crumb firming kinetics of wheat breads with anti-staling additives. *J. Cereal Sci.*, *28*, 165–174.

Asghar, A., Anjum, F. M., Ahmed, A., Hussain, S. & Tariq, M. W. (2005). Effect of polyols on quality and acceptability of frozen dough bread. *Pak. J. Food Sci.*, *15* (1-2), 11-14.

Baillet, E., Downey, G. & Tuohy, M. (2003*). Improvement of texture and volume in white bread rolls by incorporation of microbial hemicellulase preparations*, *Recent Advances in Enzymes in Grain Processing*. Laboratory of Food Chemistry, KatholiekeUniversiteit Leuven, Leuven, 255–259.

Bechtel, W. G., Meisner, D. F. & Bradley, W. B. (1953). The effect of the crust on the staling of bread. *Cereal Chem.*, *30*, 160–168.

Berglund, P. T., Shelton, D. R. & Freeman, T. P. (1991). Frozen bread dough ultra structure as affected by duration of frozen storage and freeze thaw cycles. *Cereal Chem.*, *68*, 105-107.

Bloksma, A. H. (1990), Dough structure, Dough rheology and baking quality. *Cereal Foods World*, *35*(2), 237-244.

Bollain, C., Angioloni, A. & Collar, C. (2005). Bread staling assessment of enzyme-supplemented pan breads by dynamic and static deformation measurements. *Eur. Food Res. Technol.*, *220*, 83–89.

Collar, C., Martinez, J. C., Andreu, P. & Armero, E. (2000), Effects of enzyme associations on bread dough performance. A response surface analysis. *Food Sci. Technol. Int.*, *6*(3): 217–226.

Collins, T., Gerday, C. & Feller, G. (2005). Xylanases, xylanase families and extremophilicxylanases. *FEMS Microbiology Reviews*, *29*, 3–23.

Cook, M. L. (1994). Structural Changes in the U.S. Grain and Oilseed Sector. In Lyle Chertz, P.; Lynn, M.D (ed.) *Food and Agricultural Markets: The Quiet Revolution*. USDA Economic Research Service, Washington, D.C., 118-125.

Duran, E., Leon, A., Barber, B. & Barber, C. B. (2001). Effect of low molecular weightdextrins on gelatinization and retrogradation of starch. *Eur. Food Res. and Technol.*, *212*, 203-207.

El-Dash, A. A. & Johnson, J. A. (1967). Protease enzymes: Effect on bread flavor. *Cereal Sci. Today.*, *12*, 282-288,312.

El-Hady, E. A., El-Samahy, S. K. Seibel & Brümmer, J. -M. (1996). Changes in gas production and retention in non-prefermented frozen wheat doughs. *Cereal Chem.*, *73*, 472-477.

Gerrardj. A., Newberrym, P., Rossm, J. Wilsona, E. Fayles & Kavales. (2001). Pastry lift and croissant vol-ume as affected by microbial transglutaminase. *J. Food Sci.*, *65*, 312–314.

Giannou, V., Tzia, C. & Bail., A. L. (2006). Quality and safety of frozen bakery products. In.Handbook of Frozen Food Processing and Packaging. Da-Wen Sun Ed. CRC Press, *Taylor and Francis Group FL.* 481-501.

Gray J. A. & Bemiller, J. N. (2003). *Bread Staling: Molecular Basis and Control, Comprehensive Reviews in Food Science and Food Safety*, Volume 2, Issue 1, 1–21.

Gray, J. A. & Bemiller, J. N. (2003). Bread staling: Molecular basis and control. *Compr. Rev. Food Sci. Food Safety*, *2*, 1–21.

Guy, R. (2001). Baking with Enzymes. *The World of Food Ingredients*, *26*, 28-29.

Guy, R. C. E. & Sarabjit, S. S. (2003). *Comparison of effects of xylanases with fungal amylases in five flour types, Recent Advances in Enzymes in Grain Processing.* Laboratory of Food Chemistry, KatholiekeUniversiteit Leuven, Leuven, 235–239.

Hamer, R. J. (1995). *Enzymes in the baking industry.* In: Enzymes in food processing edited by Tucker, G.A.; Woods, L.F.J.) 2ndedition. Blackie. 190-222.

Hammond, J. (1994). *Breadmaking with hemicellulase: overcoming the legal hurdles.* Food technology international Europe. Pages: 19-20, 22-23.

Haros, M., Rosell, C. M. & Benedito, C. (2002). Effect of different carbohydrates on fresh bread texture and bread staling. *Eur. Food Res. and Technol*, *215*, 425-430.

Haseborg E. & Himmelstein, A. (1988). Quality problems with high fiber breads solved by using hemicellulase enzymes. *Cereal Foods World*, *33*, 419-422.

Hille, J. D. R. & Schooneveld-Bergmans, M. E. F. (2004). Hemicellulases and Their Synergism in Breadmaking. *Cereal Food World*, *49*, 283-286.

Hino, A., Takano, H. & Tanaka, Y. (1987). New Freeze-Tolerant for Frozen Dough Preparations. *Cereal Chem.*, *64*, 269-275.

Holcomb, R. (2000). *More than Just Adding Wheat to Your Crop. In Future Farms: New Ideas for Family Farms and Rural Communities.* Poteau, Oklahoma: Kerr Center for sustainable Agriculture, Feb 8-9, 28-31.

Holmes, J. T. & Hoseney, R. C. (1987). Freezing and thawing rates and the potential of using a combination of yeast and chemical leavening. *Cereal Chem.*, *64*(4), 348-351.

Hoseney, R. C. (1998). *Principles of Cereal Science and Technology.* 2nd Ed., American Association of Cereal Chemists, St. Paul, MN.

Hudson, H. M., Daubert, C. R. & Foegeding, E. A. (2000). Rheological and physical properties of derivitized whey protein isolate powders. *J. Agric. Food Chem.*, *48*, 3112–3119.

Hozová, B., Jančovičová, J., Dodok, L., Buchtová, V. & Staruch, L. (2002). Use of transglutaminase for improvement of quality of pastry produced by frozen-dough technology. *Czech J. Food Sci.*, *20*, 215–222.

Inoue, Y. W. Bushuk. (1992). Studies on frozen doughs. II. Flour quality requirements for bread production from frozen dough. *Cereal Chem.*, *69*, 423-428.

Jiang, Z., Li, X., Yang, S., Li, L. & Tan, S. (2005). Improvement of the breadmaking quality of wheat flour by the hyperthermophilicxylanase from Thermotogamaritima. *Food Res. Int.*, 38, 37–43.

Lagendijk, J. & Pennings, H. J. (1970). Relation between complex formation of starch with monoglycerides and the firmness of bread. *Cereal Science Today*, *15*, 354-365.

Lou, J. & Wilson, W. W. (1998). *Value-added wheat products: Analysis of markets and competition*. Agricultural Economics Report No. 386, Department of Agricultural Economics, North Dakota State University.

Maarel, M. J. E. C., Veen, B. & Uitdehoog, J. C. M. (2002). Properties and applications of starch-converting enzymes of the α-amylase family. *J. of Biotechnol.*, *94*, 137-155.

Maat, J., Roza, M., Verbakel, J., Stam, H., Santosdasilva, M. J. & Bosse, M. (1992). *Xylanases and their application in bakery*. In Xylans and Xylanases.Visser, J.; Beldman, J.G.; Kustersvan, M.A.; Voragen, A.G.J. eds., Elsevier Science, Amsterdam, Netherlands. 349–360.

Martinez-Anaya, M. A. & Jimenez, T. (1997). Functionality of enzymes that hydrolyse starch and non-starch polysaccharide in bread making, *Zeitschrift fur LebensmittelUntersuchung und Forschung.*, ***205***, 209-214.

Mathewson P. R. (1998). American Association of Cereal Chemists, St. Paul, MN.

Mathewson, P. R. (1998). Enzymes. American Association of Cereal Chemists, St. Paul, MN.

Mathewson, P. R. (2000). Enzymatic activity during bread baking. *Cereal Foods World*, *45*, 98-101.

Matuda, T. G., Parra, D. F., Lugao, A. B. & Tadini, C. C. (2005). Influence of vegetable shortening and emulsifiers on the unfrozen water content and textural properties of frozen French bread dough. *Lebensmittel- Wissenschaft und –Technologie.*, *38*, 275–280.

Mercier, M. & Gélinas, P. (2001). Effect of Lipid Oxidation on Dough Bleaching. *Cereal Chem.*, *78*, 36-38.

Morita, N., Arishima, Y., Tanaka, N. & Shiotsubo, S. (1997). Utilization of Hemicellulase as Bread Improver in a Home Baker. *Journal of Applied Glycoscience*, *44*, 143-152.

Poutanen, K. (1997). Enzymes: An important tool in the improvement of the quality of cereal foods. *Trends in Food Sci. and Tech.*, *8*(9), 300-306.

Qi Si, J. (1997). Synergistic effect of enzymes for breadmaking. *Cereal Food World*, *42*, 802–807.

Resanen, J. & Laurikainen, T. (1997). Autio, K. Fermentation stability and pore size distribution of frozen prefermented lean wheat doughs. *Cereal Chem.*, *74* (1), 56-62.

Ribotta, P. D. & Le Bail, A. (2007). Effect of additives on the thermo-mechanical behaviour of dough systems at sub-freezing temperatures. *European Food Research Technology*, *224*, 519-524.

Ribotta, P. D., Leon, A. E. & Anon, M. C. (2003). Effects of Yeast Freezing in Frozen Dough. *Cereal Chem.*, *80*, 454-458.

Rogers, D. E., Day, D. D. & Olewnik, M. C. (1995). Development of an Objective Crumb-Grain Measurement. *Cereal Chem.*, *40*, 498-501.

Rouau, X., El-Hayek, M. L. & Moreau, D. (1994). Effect of an enzyme preparation containing pentosanases on the bread making quality of flours in relation to changes in pentosan properties. *J. Cereal Sci.*, *19*, 259-272.

Sahi, S. S. & Guy, R. C. E. (2004). *New Lipase Functionality in Bakery Products.* Proceedings of the 12th international ICC cereal and breadcongress, Edts: Cauvain, S. P., Salmon, S. S., and Young, L. S., Woodhead publishing Ltd.

Sato, K., Nakamura, M., Koizumi, S. & Kawachi, K. (1995). Changes in hydrophobicity and SH content on salt-induced gelation of whey protein. *J Japan Soc. Food Sci.*, *42*, 7–13.

Sato, N., Sato, M. & Nagashima, A. (1991). Effect of an enzyme preparation containing pentosanases on the breadmaking quality of flours in relation to changes in pentosan properties. *J. Cereal Sci.*, *19*, 259–272.

Schroeder, E. (1999). *Frozen Dough: In the Footsteps of Fresh Baked.* Milling & Baking News.

Schultz, A. S., Schoonover, F.D., Fisher, R. A. & Jacker, S. S. (1952). Retardation of crumb starch staling in commercial bread by bacterial α-amylase. *Cereal Chem.*, *29*, 200-207.

Sluimer, P. (2005). *Principles of Breadmaking: Functionality of Raw Materials and Process Steps.* American Association of Cereal Chemists, St. Paul, MN.

Stampfli, l. & Nersten, B. (1995). Emulsifiers in breadmaking-a review. *Food Chem.*, *52*, 353-360.

Turgeon, S. L., Gauthier, S. F. & Paquin, R. (1992), Emulsifying property of whey peptide fractions as a function of pH and ionic strength. *J. Food Sci.*, **57** **(3),** 601–604, 634.

U.S. Patent No. 4,990,343. (1991). *Enzyme product and method of improving the properties of dough and the quality of bread.*

Underkofler, L. A. (1972). *Chapter 1 Enzymes in Handbook of Food Additives.* 2nd Edition. Vol. *I*. CRC Press.

Valjakka, T. T., Ponte, J. G. & Kulp, K. (1994). Studies on a raw-starch digesting enzyme. I. Comparison to fungal and bacterial enzymes and an emulsifier in white pan bread. *Cereal Chem.*, *71*, 139-144.

Van Dam, H. W. & Hille, J. D. R. (1992. Yeast and enzymes in breadmaking. *Cereal Foods World.*, 37(3), 245-252.

Varriano, M. E., Hsu, K. H. & Mahdi, J. (1980). Rheological and structural changes in frozen dough. *Baker's Digest*, *54*(1), 32.

Wolt, M. J. & D'appolonia, B. L. (1984). Factors involved in the stability of frozen dough. I. The influence of yeast-reducing compounds on frozen-dough stability. *Cereal Chem.*, *61*, 209-212.

Yousif, A. K. (1998). Concentration of acidic whey and its functionality in French type bread. *Intl. J. Dairy Tech.*, *51*(3), 72-76.

INDEX

F

G

H

I

J

K

L

M

N

O

P

Q

R

S

T

U

V

W

X

Y

Z